Regenerative Medicine and Human Genetic Modification

By Edward W. Gaskin
www.amazon.com/author/ed-gaskin

Please consider leaving a review wherever you bought the book or telling your friends know about it. Help spread the word.

Thank you for purchasing and reading my book. I would love to hear your comments and thoughts on *Regenerative Medicine and Human Genetic Modification* and find out if it helped.

Regenerative Medicine and

Human Genetic Modification

Ed Gaskin

2014, Revised 2017

Table of Contents

Introduction

Regenerative medicine has profound potential to change the trajectory of medicine. Regenerative medicine is the real-life story of humanity's quest for anti-aging solutions that neither science nor fantasy writer could imagine. It has been years in the making as advances in biotechnology converge. The digital revolution brought with it the ability to disrupt one industry after another and to see many things that were separate converge. Simply look at your smart phone that can be a clock, stop watch, timer, voice recorder, camera, video recorder, phone, pocket computer running numerous programs, radio, TV, movie screen, book, calendar, journal; some phones have over 1 billion applications. This does not include how the Internet itself has changed society.

Now imagine the same revolutions taking place in biotechnology. Instead of programing on a two-digit code of zeros and ones, we can do it with a four-digit code, A, C, T, and G. We have the ability to program life at the genetic level. Just as we write code by arranging zeros and ones, we can arrange genes or the genetic code anyway we want. Some call this genetic programing or programing with genes and enables genetic modification. Because genes are the building blocks of life, we are not limited to using genes from within species, e.g. plants, insects, animals and humans, but among them. An example of this is BioSteel:

BioSteel was a trademark name for a high-strength based fiber material made of the recombinant spider silk-like protein extracted from the milk of transgenic goats, made by Nexia Biotechnologies, and later by the Randy Lewis lab of the University of Wyoming and Utah State University.[1] It is reportedly 7-10 times as strong as steel if compared for the same weight, and can stretch up to 20 times its

unaltered size without losing its strength properties. It also has very high resistance to extreme temperatures, not losing any of its properties within -20 to 330 degrees Celsius.
From Wikipedia

BioSteel is the result of using technology to **combine genes from a spider with that of a goat.** With this type of material properties, there are a range of amazing applications including artificial limbs, bulletproof vests, and parachutes. These types of developments are called "transgenetic" as they are not limited to the genes of one species.

While these technologies hold great promise, they also bring with them tremendous peril as they enable us to modify human life at the genetic level. Meaning, once we intervene at the genetic level, that genetic change can be passed down to future generations. The tools necessary to create new tissues and organs, the basis for regenerative medicine are also the same tools necessary to genetically modify humans. The ability to genetically modify humans may also provide amazing cures to chronic, genetic and infectious diseases such as HIV. At the moment, as a society, we have not weighed in as to what we believe is acceptable and what is not when it comes to genetic modification of humans. Do the benefits outweigh the risks, and what should be the limits if any?

We are no longer debating the tools and their sources, but their future application. The tools and technology that makes regenerative medicine possible also make human genetic modification possible. We need to carefully work our way through this debate so we set the limits in such a way that we don't inhibit medicine and healing and at the same time protect human worth and dignity. We have the ability to genetically modify humans using novel gene sequences developed by computers, gene sequences that have never existed before in nature. This is similar to developing novel pharmaceuticals in a lab.

The core question of the book in a nutshell is:

Based on ethical considerations, what if any boundaries should exist when it comes to the genetic modification of humans?

According to Wikipedia, the history of regenerative medicine starts with Geron Corporation. From 1995 to 1998, Michael D. West, PhD organized and managed the research between Geron Corporation and its academic collaborators James Thomson at the University of Wisconsin-Madison and John Gearhart of Johns Hopkins University that led to the first isolation of human embryonic stem and human embryonic germ cells. While he did not coin the term "regenerative medicine" Dr. West is recognized as its founder. Because Dr. West is the father of regenerative medicine, a recognized thought leader (based on his published research and speaking in this area), as well as having led three separate biotech companies doing work in the field of regenerative medicine, we focus on his work and thinking in this book.

We last seriously considered these issues as a nation, when we decided that reproductive cloning was not acceptable but therapeutic cloning was. We said that experimenting on human embryos prior to implantation or 14 days when the primitive streak is formed is ok, but not afterward. While we did say that it was not ok to use these technologies to create a human, we did not say it wasn't ok to modify the genetic structure found in an embryo and then grow it to term. In other words, ***we did not say we can't use the tools*** *of genetic engineering, gene targeting, genomics, cloning, embryonic stem cells, etc. to modify a human being. And if we do so, what are the limits?*

Developing these ethical guidelines is a challenge we must address and address quickly. In this book, we will not only look at the history of how we developed the ethical guidelines that make regenerative medicine possible but also demonstrate that ethical guidelines are needed to guide us on the topic of genetic modification of humans.

In 2014, two cases were reported of genetically modifying humans. One involved helping a woman with infertility to have children.[1] The other had to do with helping a child with mitochondrial disease.[2] Since then the Chinese have genetically modified two human embryos.[3]T They were trying to genetically modify embryos to be resistant to HIV.

With all of the debate on embryonic stem cells versus adult stem cells, and therapeutic versus reproductive cloning, the scientific community heard loud and clear there was an uneasiness in using embryonic stem cells, even if there was an ethically scientific reason to do so. As a result, scientists discovered new ways to obtain embryonic stem cells without the destruction of human embryos.

---

[1] World's first GM babies born by Michael Hanlon, Daily Mail, http://www.dailymail.co.uk/news/article-43767/Worlds-GM-babies-born.html.

[2] Genetically Modified Babies, By Marcy Darnovsky, Feb. 23, 2014.

[3] Chinese scientists have genetically modified a human embryo

AGAIN, Science Alert April 11, 2016.

One technique was to use a parthenogenetic process. According to the website "Fight Aging":

> Parthenogenesis is a form of asexual reproduction that occurs naturally in plants, insects, fish, amphibians and reptiles. During this process, unfertilized eggs begin to develop as if they've been fertilized. In 2007, researchers induced human egg cells with chemicals mimicking fertilization so they would undergo the process. The results were parthenogenetic cells that share the same properties as embryos, except that they can't grow further. The cells are akin to pluripotent stem cells derived from embryos, which means they have the ability to develop into different types of cells - including heart cells.
>
> [Researchers] used this knowledge to turn body cells of mice into parthenogenetic stem cells, which were then grown into mature, functional cardiomyocytes. Researchers used these cells to engineer myocardium - heart muscle - with the same structure and function of normal myocardium. The muscle was then grafted onto the hearts of the mice that had contributed the original eggs for parthenogenesis, where it worked the same way as existing muscle.

Another option is to reprogram adult stem cells to turn them into embryonic stem cells, which is called "induced pluripotent stem cells." According to Wikipedia:

> **Induced pluripotent stem cells** (also known as **iPS** cells or **iPSCs**) are a type of pluripotent stem cell that can be generated directly from adult cells. Since iPSCs can be derived directly from adult

tissues, they not only bypass the need for embryos, but can be made in a patient-matched manner, which means that each individual could have their own pluripotent stem cell line. These unlimited supplies of autologous cells could be used to generate transplants without the risk of immune rejection. While the iPSC technology has not yet advanced to a stage where therapeutic transplants have been deemed safe, iPSCs are readily being used in personalized drug discovery efforts and understanding the patient-specific basis of disease.

These two techniques made the former arguments almost moot as it was now possible to obtain embryonic stem cells without the destruction of human embryos. And there are other potential ways to obtain embryonic stem cells. Science probably would not have made these discoveries unless they had the moral opposition to the use of human embryos. What we are arguing here is the need for the public to express their beliefs on human genetic modification so those in government and in the scientific community have a feeling about what we want in terms of what should and should not be allowed. As we get closer to genetically modifying human beings, we should have Congressional hearings like we did for embryonic stem cells and cloning so there is a record of how and why we made the decisions we did.

The four technologies - embryonic stem cells, nuclear transfer, genetic engineering, and genomics – each hold both great promise in medicine and great peril. In addition to the risk each individually poses, when combined they provide the promise of regenerative medicine and the promise and danger of genetic modification. These four technologies turn the genetic code of insects, plants, animals and humans into biological "Tinker Toys" waiting to be assembled in any way we desire or allow. We can simply mix and match them based on our desires. Just as there are "hacks" for everything from games to toys to software code, there could be "hacks" with biological organisms, including human organisms.

The developments in the area of genetic programing or genetic modification will occur faster than they did in the Internet age, as inventors can leverage much of the learning already done in the software area. For instance, one can imagine libraries of genetic code from every species imaginable, ready to be bought and plugged in at a moment's notice, similarly to the purchase and use of software code, objects, and tools. This blend of high tech and biotech will have no shortage of funding as investors look to create the biotech equivalent of Amazon, Facebook, and Google.

At the Woodrow Wilson International Center for Scholars, they had a Director's Forum panel on the stem cell controversy with Dr. William Hurlbut, Dr. Robin Cook and Dr. Michael West on January 18, 2006. Dr. West said while this new technology has tremendous upside, he is troubled by human genetic modification and other such applications of these new technologies. So I wrote to him to ask:

Does BioTime have any guidelines regarding the genetic modification of humans in terms of what is and what is not ethically acceptable?

Dr. West's response was:

"In regard to genetic engineering of human life, there are a gazillion number of such modifications one could consider. Some would be intended to be of a beneficent nature (like correcting the sickle cell gene), others with a more hubristic theme like double muscling. I suggest that making 'super people' should not, in principle, be condemned; it's just that uses other than correcting disease genes should have extremely high standards of safety."

If one of the pioneers of the field, who is closest to the problem and implications does not have any guidelines, perhaps no one else does either and therefore it is worth discussing. The call for hearings on the topic or at least a set of recommendations from the Presidential Commission for the Study of Bioethical Issues should be made.

This book was written to give us a look at what got us here by looking at the arguments around therapeutic cloning and embryonic stem cells, and where we are headed with the hopes it will further discussion and the development of ethical guidelines in this area.

Important Background and Updates[4]

When people believed the world was flat, they did not try to sail around the world. If you believe getting old and dying is like the law of gravity, then you don't try to change it. You take it as a given. Those who try to develop anti-aging techniques are thought of in the same vein as those who are trying to develop anti-gravity or perpetual motion machines. However, now there are serious scientists doing research in this area including Google. Google has funded a company called, Cailco with the goal of fighting aging and age related diseases and has committed to investing hundreds of millions of dollars into the company. In fact, Cailco has formed significant partnerships in this area.[5]

There are lots of strategies for trying to stop, slow or reverse aging. See Wikipedia under "Anti-aging movement" and "Life Extension" research. Anti-aging strategies cover a range from diets and supplements, hormone treatments, caloric restrictions, anti-aging drugs, and cryonics among others. Many of these approaches have

---

[4] Since the first edition was published in 2014, so much has happened in terms of gene editing, specifically, CRISPR-Cas9 and human genetic modification, in particular the International Summit on Human Gene Editing, that I felt a shot update was needed. I kept the chapter on the Primitive Streak because the ethics of experimentation on pre implantation embryos is still relevant in this age of genetic modification.

[5] Far from being quack scientists, Google's Calico has brought in Arthur Levinson CEO of Genentech and in September of 2015, Google disclosed that it gave $240 million to Calico, in exchange for stock plus promised support of up $490 million should Calico need it. See Wikipedia, Calico Company. What's He Building in There? The Stealth Attempt to Defeat Aging at Google's Calico. Recode.com, Mark Bergen, December 28, 2015,

been strongly criticized by the American Medical Association. However, there is one field that has shown significant promise and that is looking at aging at the genetic level, **molecular gerontology**.

When we look at aging related diseases such as Hutchinson Gilford Progeria syndrome, and Wieners syndrome, which would indicate that aging is genetic as opposed to a result of wear and tear. We understand the genetics of these diseases fairly profoundly. Also in nature, some organisms have very long life spans and some are even believed to be biologically immortal. Such examples include:

- Rough eye rockfish - 205 years
- Red sea urchin -   200 years
- Ocean quahog clam - 507 years

This would seem to indicate that the biological clock is moving slower in some species than others. The conclusion is, the best "anti-aging strategy is to try to develop interventions at the genetic or molecular level.

There is a history of studying aging at the cellular level. In an over simplification, the various cells that make up all of our biological systems get old and die or "degenerate" band new ones to replace the old ones become fewer and fewer, leading to a wide range of age related diseases.

This brings us to **Regenerative Medicine**, defined in a Wikipedia article as the "process of replacing, engineering or regenerating human cells, tissues or organs to restore or establish normal function".[2] The function is often lost due to age, disease, damage, or congenital defects. This field holds the promise of engineering damaged tissues and organs by stimulating the body's own repair mechanisms to functionally heal previously irreparable tissues or organs.[3]

In the history of molecular gerontology, the idea of extending the life of a cell has been explored as well as making old cells new again. What regenerative medicine does is simply provide new cells, or tissues or organs made from young healthy cells in various therapeutic applications.

There are still Model T automobiles on the road today, and they could still be on the road 100 or 200 years from now, as long as whatever broke on them could be replaced or repaired. In the same way, humans could live for much longer if what was broken or damaged could be repaired. Regenerative medicine holds such promise.[6]
Those in the regenerative medicine field see applications in various categories.

Rejuvenation – using stem cells to support the body's natural ability to heal itself. For example, after a cut, your skin heals itself within a few days. The idea would be to replicate this process in an effort to help the body repaid other types of damage.

Replacement -using healthy cells, tissues or organs to replace damaged ones.

From the NIH website Regenerative Medicine, they list the possible applications of this game changing technology.

- Imagine a world where there is no donor organ shortage, where victims of spinal cord injuries can walk, and where weakened hearts are replaced. This is the long-term promise of regenerative medicine, a rapidly developing field with the potential to transform the treatment of human disease through

---

[6] Wikepedia has a great article on the history of regenerative medicine, so I won't repeat it here.

the development of innovative new therapies that offer a faster, more complete recovery with significantly fewer side effects or risk of complications.

- Insulin-producing pancreatic islets could be regenerated in the body or grown in the laboratory and implanted, creating the potential for a cure for diabetes.

- Tissue-engineered heart muscle may be available to repair human hearts damaged by attack or disease.

- The emerging technique of Organ Printing utilizes a standard ink jet printer modified with tissue matrix material (and possibly also cells) replacing the ink. "Made-to-order" organs of almost any configuration could then be cast and implanted.

- Materials Science meets Regenerative Medicine as "smart" biomaterials are being made that actively participate in, and orchestrate, the formation of functional tissue.

- New approaches to revitalizing worn-out body parts include removing all of the cells from an organ, and infusing new cells to integrate into the existing matrix and restore full functionality.

The future of regenerative medicine is called **Induced Tissue Regeneration (iTR™)**[7]. Meaning humans would have the ability to control the regeneration of tissues throughout life. The human body is powerful enough to

---

[7] **Induced Tissue Regeneration** – The induction of the ability to recapitulate embryogenesis in the tissue of an adult mammal by means of the administration of exogenous reprogramming factors in vitro or in vivo.

organize itself from the embryo but lacks the ability to maintain what it has created. Why? The human body starts at the embryo stage and early on has the ability to regenerate or rebuild many body parts, tissues. So a cut will heal without a scar. Over time, the body loses such potential. This switch is thought to prevent tumors or cancer. What if the natural regenerative potential of the human body was maintained longer in life or throughout life? The basic idea is some organisms such as the Mexican Salamander have very powerful ability to regenerate limbs and other body parts throughout their life. The Mexican Salamander grows a new limb exactly like the old one. What if humans had that same capability? So if we had a heart attack and needed a new heart, our body would grow one, or if in an industrial accident we lost a limb, we could grow a new one.

Regenerative medicine requires the use of stem cells. In prior discussion, regenerative medicine was bogged down in abortion or "Prolife" ethics. People often debated the use of embryonic versus adult stem cells, because embryonic stem cells did require the destruction of an embryo or the taking of a "life." In the first version of this book, a great deal of time was spent dealing with the ethics of creating embryonic stem cells. What was particularly appealing about embryonic stem cells is that they are pluripotent meaning they could become any type of cell. Shinya Yamanaka, a Japanese scientist pioneered work in what is called **induced pluripotent stem cells,** which did not require the destruction of a human embryo and thus avoided the ethical debate all together. Now pluripotent stem cells could be made from adult skin cells. In addition, induced pluripotent stem cells can be made from the patient's own genetic material, so matching is not a problem. Science has advanced in such a way to negate the discussion or use of pre-implantation embryos altogether, making most of the prior debates moot. In other words, we can use tissue from any part of the body to create embryonic stem cells, and it no longer requires the destruction of a human embryo.

Stem cells are the foundation of regenerative medicine. Imagine you had one "magic" seed or seed type, and from that seed, you could direct or program it to become any type of known plant, flower, tree that existed. Imagine a beautiful botanical garden and every plant you saw coming from the same seed. Embryonic stem cells are pluripotent and are similar to the magic seed we just described. Pluripotent cells can give rise to all of the cell types that make up the body. Using the magic seed example pluripotent stem cells allow us to create any type of plant or cells or tissue or organ in the human body. In the case of the Model T example, we can now create any part we need to fix or repair any damage to our car, or body should it become damaged. Because we are treating the problem at the root cause or at the cellular level, we are no longer treating the symptoms.

Regenerative medicine often uses cloning to create stem cells. Because these cells are from the patient, the resulting cells, tissue or organs won't be rejected. By making the distinction between "Reproductive" i.e. using cloning to reproduce copies of humans versus "Therapeutic" i.e. using cloning for therapeutic or medical purposes, the debate on the use of cloning has also dissipated even though there is no national probation against human cloning in the US.

Regenerative medicine is possible because embryonic stem cells can be engineered or programed or reprogramed to become any cell type in the human body. This combined with new advances in gene editing has opened the door to a whole new world of therapies.

What this means is we will have the ability to create any new tissue or organ we need. But with the power of gene editing, we can engineer or "tweak" cells, tissues or organs. For example, we know what genetically, what makes someone immune to HIV. We could create for them a new immune system that had that characteristic and thus make them immune to HIV. If someone had a genetic defect that affected one of their organs, we could replace that organ without the defect. We could build them a more powerful heart than what they were genetically programmed to have.

CRISPR-Cas9 and Gene Editing

Gene editing is similar to computer programing. Computer programing at its base level is based on a series of zeroes and ones. From that you can create any type of computer application in the world. Genetic programming is similar but based on a series of A, C, T, and G. A, C, T and G and called bases and are what make up DNA, which makes up our genes. By changing the order of these in the human genome, you can change any and every part of a person, from a change to an organ to a change in height and other possible dimensions. What makes gene editing faster, easier and more accurate than ever before is the development of Clustered Regularly Interspaced Short Palindromic Repeats or CRISPR-Cas9.

A good overview of CRISPR-cas9 can be found at, **"What is CRISPR-Cas9?**[8] But what CRISPR does is make it much easier to change, rearrange DNA and RNA. Not only does this have applications with regenerative medicine, but also with gene therapy. Another overview of CRISPR its

---

[8] http://www.yourgenome.org/facts/what-is-crispr-cas9

history and applications is **"Easy DNA Editing Will Remake the World. Buckle Up,"[9]**

For decades we have talked about designer babies or genetic engineering. But now all of the pieces are in place. One can get a copy of our own or anyone's human genome. The cost of sequencing someone's genome has dropped from $2.7 billion and took 15 years, to less than $1,000 today. We now know more about genetic mutations and defects and their relationship to disease and how to fix them. CRISPR provides us the tool to fix or delete genetic defects. Therapeutic cloning is a readily available tool, and we have the ability to create induced pluripotent stem cells. As a result of these and other advances, we have very powerful tools at our disposal.

International Summit on Human Gene Editing

The International Summit on Human Gene Editing December 1-3, Washing, D.C. was organized by the US national academies of sciences and medicine, the Royal Society in London and the Chinese Academy of Sciences (CAS). The meeting highlighted China's emerging prominence in genomics; much of the discussion surrounded an April publication by Chinese researchers who used the gene-editing technology CRISPR–Cas9 to modify a gene in non-viable human embryos (P. Liang *et al. Protein Cell* **6,** 363–372; 2015).[10]

The **International Summit on Human Gene Editing** still has an active homepage that can be found at http://nationalacademies.org/gene-editing/index.htm. The

---

[9] Wired, July 22, 2015.

[10] Nature.com

website will provide presentation materials from the conference including slides and in some cases videos. A summary of the conference can be downloaded here. ttps://www.nap.edu/catalog/21913/international-summit-on-human-gene-editing-a-global-discussion.

Other summaries of the conference can be found at:

- **Nature.com** - Global summit reveals divergent views on human gene editing

- **The National Center for Biotechnology Information** - International Summit on Human Gene Editing: A Global Discussion, Meeting in Brief.

The following is an excerpt from the National Academy Press's eight page final book on the conference.[11] The following is the conferences closing statement:

On Human Gene Editing: International Summit Statement by the Organizing Committee
Scientific advances in molecular biology over the past 50 years have produced remarkable progress in medicine. Some of these advances have also raised important ethical and societal issues – for example, about the use of recombinant DNA technologies or embryonic stem cells. The scientific community has consistently recognized its responsibility to identify and confront these issues. In these cases, engagement by a range of stakeholders has led to solutions that have made it possible to obtain major

---

[11] "International Summit on Human Gene Editing A Global Discussion." National Academies of Sciences, Engineering, and Medicine. 2016. *International Summit on Human Gene Editing: A Global Discussion.* Washington, DC: The National Academies Press. doi: 10.17226/21913.

benefits for human health while appropriately addressing societal issues.

Fundamental research into the ways by which bacteria defend themselves against viruses has recently led to the development of powerful new techniques that make it possible to perform gene editing – that is, precisely altering genetic sequences – in living cells, including those of humans, at much higher accuracy and efficiency than ever before possible. These techniques are already in broad use in biomedical research. They may also enable wide-ranging clinical applications in medicine. At the same time, the prospect of human genome editing raises many important scientific, ethical, and societal questions.

After three days of thoughtful discussion of these issues, the members of the Organizing Committee for the International Summit on Human Gene Editing have reached the following conclusions:

1. **Basic and Preclinical Research**. Intensive basic and preclinical research is clearly needed and should proceed, subject to appropriate legal and ethical rules and oversight, on (i) technologies for editing genetic sequences in human cells, (ii) the potential benefits and risks of proposed clinical uses, and (iii) understanding the biology of human embryos and germline cells. If, in the process of research, early human embryos or germline cells undergo gene editing, the modified cells should not be used to establish a pregnancy.

2. **Clinical Use Somatic**. Many promising and valuable clinical applications of gene editing are directed at altering genetic sequences only in somatic cells – that is, cells whose genomes are not transmitted to the next generation. Examples that

have been proposed include editing genes for sickle-cell anemia in blood cells or for improving the ability of immune cells to target cancer. There is a need to understand the risks, such as inaccurate editing, and the potential benefits of each proposed genetic modification. Because proposed clinical uses are intended to affect only the individual who receives them, they can be appropriately and rigorously evaluated within existing and evolving regulatory frameworks for gene therapy, and regulators can weigh risks and potential benefits in approving clinical trials and therapies.

3. **Clinical Use: Germline.** Gene editing might also be used, in principle, to make genetic alterations in gametes or embryos, which will be carried by all of the cells of a resulting child and will be passed on to subsequent generations as part of the human gene pool. Examples that have been proposed range from avoidance of severe inherited diseases to 'enhancement' of human capabilities. Such modifications of human genomes might include the introduction of naturally occurring variants or totally novel genetic changes thought to be beneficial.

Germline editing poses many important issues, including: (i) the risks of inaccurate editing (such as off-target mutations) and incomplete editing of the cells of early-stage embryos (mosaicism); (ii) the difficulty of predicting harmful effects that genetic changes may have under the wide range of circumstances experienced by the human population, including interactions with other genetic variants and with the environment; (iii) the obligation to consider implications for both the individual and the future generations who will carry the genetic alterations; (iv) the fact that, once introduced into the human population, genetic alterations would be difficult to remove and would not remain within any single community or country; (v) the possibility that permanent genetic 'enhancements' to subsets of the population could exacerbate social inequities or be used coercively; and (vi) the moral and ethical considerations in purposefully altering human evolution using this technology.

It would be irresponsible to proceed with any clinical use of germline editing unless and until (i) the relevant safety and efficacy issues have been resolved, based on appropriate understanding and balancing of risks, potential benefits, and alternatives, and (ii) there is broad societal consensus about the appropriateness of the proposed application. Moreover, any clinical use should proceed only under appropriate regulatory oversight. At present, these criteria have not been met for any proposed clinical use: the safety issues have not yet been adequately explored; the cases of most compelling benefit are limited; and many nations have legislative or regulatory bans on germline modification. However, as scientific knowledge advances and societal views evolve, the clinical use of germline editing should be revisited on a regular basis.

4. **Need for an Ongoing Forum**. While each nation ultimately has the authority to regulate activities under its jurisdiction, the human genome is shared among all nations. The international community should strive to establish norms concerning

acceptable uses of human germline editing and to harmonize regulations, in order to discourage unacceptable activities while advancing human health and welfare.

We therefore call upon the national academies that co-hosted the summit – the U.S. National Academy of Sciences and U.S. National Academy of Medicine; the Royal Society; and the Chinese Academy of Sciences – to take the lead in creating an ongoing international forum to discuss potential clinical uses of gene editing; help inform decisions by national policymakers and others; formulate recommendations and guidelines; and promote coordination among nations.

The forum should be inclusive among nations and engage a wide range of perspectives and expertise – including from biomedical scientists, social scientists, ethicists, health care providers, patients and their families, people with disabilities, policymakers, regulators, research funders, faith leaders, public interest advocates, industry representatives, and members of the general public.

Statement by

Ralph J. Cicerone, President, U.S. National Academy of Sciences
Victor J. Dzau, President, U.S. National Academy of Medicine
Chunli Bai, President, Chinese Academy of Sciences
Venki Ramakrishnan, President, The Royal Society

Introduction to Chapter on Primitive Streak

In the living room of Dr. West, over 20 years ago, Dr. West asked friends a hypothetical question, "If there was a pill that would extend your life, would you take it?" My first thought, which led to my counter question was, "If such a pill existed, how much would it cost and who would control access to it?" Dr. West had been asking that question of friends for weeks. Something was on his mind.

Mike looked at Progeria, a rare genetic condition that produces rapid aging in children. He hypothesized that if aging could be accelerated, it could be slowed down; off he went to do just that. Was this another of history's fools in search of the "Fountain of Youth?" No, he was a man with a vision and passion. On the way he made a discovery that was 2,500 times more potent than Retin-A, the wrinkle compound. Instead of stopping there, to commercialize that discovery, he kept going because he said, "no one ever died of wrinkles or baldness." He discovered how telomeres worked, and, while it did have a direct impact on aging, slowing it down or reversing it was not going to be as easy as he thought. Based on an observation about tadpoles, he thought that embryonic stem cells could lead to what he would later call "regenerative medicine" and the rest was history. (You can read a richer history in his book, *The Immortal Cell*, and in a *Fortune* article[12]) After founding Geron Corporation, Origen Therapeutics Inc., and Advanced Cell Technologies, he now runs BioTime Inc. He is rightly referred to as the father of Regenerative Medicine.

Before Dr. West started his quest, he said,

---

[12] Who Is Doctor West, And Why Has He Got George Bush So Ticked Off?". *Fortune Small Business*. April 1, 2002.

"I went to see Francis Schaeffer while he was undergoing treatment at the Mayo clinic in Rochester MN. On that visit, I asked him, if I had a pill that would keep you alive forever, would you take it? And he said yes."

He took the position that many Biblical scholars did, that it was our job to reverse the curse of the fall, and sickness and death was one of the effects of the fall.

As Dr. West started researching and advocating for the use of embryonic stem cells and therapeutic cloning, people would challenge his Christian beliefs and values.

Dr. West was an Evangelical Christian who had started a Christian coffee house so Christians would have an alternative to bars and cubs. At the coffee house, in addition to Christian singers, he had Bible studies and classes. He even brought in a professor to teach Hebrew so Christians could learn to read the Bible in the original languages. Mike hated liberal Christianity. He attended a Bible church that was part of the IFCA or Independent Fundamentalist Churches of America denomination. He bought the Francis Schaeffer series on "What Ever Happened to the Human Race" and taught that series as well as took a group of Christian to see the "How Should we Then Live" Series also by Francis Schaeffer. Needless to say, he was Pro Life and very familiar with the Biblical and ethical arguments as he explored the merits of stem cell research and therapeutic cloning.

This looks like a clear conflict, science versus religion.[13] Dr. West was familiar with this type of conflict. Early in

---

[13] If determining if life begins is based on when ensoulment occurs, then it is not really a matter of scientific inquiry.

his career, he studied at Andrews University which believed in a "young earth" as opposed to the traditional Old Earth Theory as it related to the Biblical view of creationism.

However, what if there was a seemingly easy solution to this impasse? What if one understood conception to take place at the time individualization, approximately two weeks after fertilization, not at the moment of fertilization? If this interpretation were plausible, then the impasse could be solved. The two week milestone was not arbitrary, but it is when the Primitive Streak[14] occurs.

If this interpretation of conception was accepted, Christians could still maintain a conservative, literal understanding of the Bible regarding the passages of scripture that state life starts at conception and that life begins in the womb.

This is the view of Dr. Norman Ford, a Salesian priest and moral philosopher, takes. He concludes that we did not begin before definitive individuation, which occurs with the appearance of the primitive streak at 14 days after fertilization. Dr. Ford was exploring the issue because of various reproductive technologies and the issue of

---

[14] The primitive streak is a structure that forms during the early stages of embryonic development. The primitive streak is an important concept in bioethics, where some experts have argued that experimentation with human embryos is permissible, but only before the primitive streak develops, generally around the fourteenth day of existence. (The President's Council on Bioethics, Human Cloning and Human Dignity: An Ethical Inquiry. July 2002). The development of the primitive streak is taken, by such bioethicists, to signify the creation of a unique, human being. In some countries, it is illegal to develop a human embryo for more than 14 days outside a woman's body. Prohibition of Human Cloning for Reproduction Act 2002". Government of Australia Department of Health and Ageing. 22 Dec 2008.

ensoulment, or "animation" – when a person gets a soul – which has been long debated in Church history. The idea is simply you don't get a soul until you are a person. Having an abortion for example prior to the embryo becoming ensouled was considered acceptable because no human life was destroyed.

When Christians have used the Bible to make the case the destruction of the embryo is wrong, they often point to four types of passages. One is Exodus 21:22-25. Since children are a blessing from God, the loss of such blessing should be compensated. This demonstrates that life in the womb had value. The second type is when sex occurs outside of marriage, e.g. fornication, adultery and a pregnancy occurs. In this case people have an abortion to cover up the other sin, which only compounds the sins. The third type of scriptural references is all of the ones that mention the word conception and life beginning in the womb. None of these verses provide insight as to when conception occurs, e.g. at fertilization or individualization. The fourth type of passage is concerning stories where a woman is barren and prays to God for a child. God answers her prayer with the blessing of a child. What is implied in this case is that conception is controlled by God and therefore willfully killing what God creates is wrong. Those who take that position argue the scientific argument, the church's teaching on ensoulment or animation is irrelevant, because only God can create life.

The question is this: has the church always said that life begins at conception where conception equals fertilization? The answer is that there has been debate within the church. Some have argued human life does not occur at conception but at some other point, e.g. when the rational soul enters the body, when sin is transmitted, when "animation" takes place, or when the quickening occurs, or when a baby takes its first breath, among other views.

The position that life begins at some other point is referred to as "hominization" - the evolutionary development of human characteristics that differentiate hominids from their primate ancestors (Merriam Webster). Some in the Catholic Church taught delayed hominization.

The debate on when life begins starts in 100 A.D:

One of the earliest church documents, the Didache, condemns abortion but asks two critical questions: 1) Is abortion being used to conceal the sins of fornication and adultery? and 2) Does the fetus have a rational soul from the moment of conception, or does it become an ensouled human at a later point?[15]

The "delayed ensoulment" belief of Aristotle (384-322 BCE) was widely accepted in Pagan Greece and Rome. He taught a fetus originally has a vegetable soul. This evolves into an animal soul later in gestation. Finally the fetus is "animated" with a human soul. This ensoulment was believed to occur at 40 days after conception for male fetuses, and 90 days after conception for female fetuses.[16] The difference was of little consequence because, in those days, the gender of a fetus could not be determined visually until about 90 days from conception, and no genetic tests existed to determine gender. Thus contraception and abortion were not condemned if performed early in gestation. It is only if the abortion is done later in pregnancy that a human soul is destroyed[17]

---

[15] Abortion and Catholic Thought: The Little-Told History, Autumn 1996 issue of Conscience This is updated, condensed and adapted from CFFC's publication The History of Abortion in the Catholic Church: The Untold Story by Jane Hurst, 1989.
[16] Aristotle: History of Animals, book III.
[17] By Tom Longua Abortion, Ancient Christian Beliefs.

In 1312, delayed hominization was confirmed.

The Council of Vienne confirmed the conception of man put forth by St. Thomas Aquinas. While Aquinas had opposed abortion -- as a form of contraception and a sin against marriage -- he had maintained that the sin in abortion was not homicide unless the fetus was ensouled, and thus a human being. Aquinas had said the fetus is first endowed with a vegetative soul, then an animal soul, and then -- when its body is developed -- a rational soul. This theory of delayed hominization is the most consistent thread throughout church's history on abortion.[18]

Pro Life and Pro Choice Catholics have debated this point more thoroughly. One of the best argued cases that life does not begin at the moment of fertilization is made by Dr. Norman Ford who wrote "The Prenatal Person: Ethics from Conception to Birth," and "When Did I Begin? Conception of the Human Individual in History, Philosophy, and Science."

Darryl R. J. Macer wrote a review of Dr. Ford's book and summarized the key argument this way:

There are several problems with placing the beginning of the individual at fertilisation, including the difference between genetic and ontological individuality, identical twinning occurs between 7-10 days later, the 70% natural embryo wastage before implantation is complete (14 days), the totipotency of early embryonic cells, the lack of unity of the cells in the early embryo, the possibility of chimeras (individuals from multiple embryonic cells) being formed,

---

[18] Joseph F. Donceel, S.J., "Immediate Animation and Delayed Hominization," Theological Studies, vols. 1 & 2 (New York: Columbia University Press, 1970), pp. 86-88.

and recombination (two embryos combine to form one), and parthenogenesis where the embryo is not the result of a fertilised egg with the new genotype, the possibility of a cancerous tumour being the outcome of embryonic development. There are important philosophical problems with ensoulment occurring before an individual exists.

Implantation is the next major stage (7-14 days), and it has some significance for the stability that is occurring. More significant is the formation of the primitive streak at 14 days that makes a beginning of the clump of cells becoming an individual coordinated embryo. By 3 weeks the process called gastrulation is completed where the embryo has formed the three basic types of tissue and the membranes around the embryo are well underway. Ford concludes that the time of individualisation is 14 days, the time from which we began. There is some logic in saying that a "human individual could scarcely exist before a definitive human body is formed", fertilisation is to be considered as the beginning of the development into a human individual.

Based on Dr. Ford's position as a Catholic who has reviewed both the church's position and the biological evidence, you can see it is similar to the position taken in the next article. That position is the formation of the Primitive Streak should be the ethical dividing line, and is consistent with an understanding of life beginning at conception, if one interprets conception as individualization, not fertilization. The only other choice is to start at the three-week market when gastrulation is completed and where the embryo has formed the three basic types of tissue.

Karl Ernest von Baer discovered the ovum in the female in 1827, and by 1875 the joint action of spermatozoon and ovum in generation had been determined. Making an adequately informed biological argument earlier would have been impossible, unless what was occurring at the cellular level was not what mattered. [19] This changes the discussion away from one of science versus religion and if the Bible is a book of science or just contains science, or if one is Pro Life or Pro Choice, it becomes an issue of interpretation.

In history we have seen that philosophy influences theology and culture influences our interpretation of the Bible. We can also see that our scientific understanding and world view have an influence on our interpretation of certain passages as well. Over the centuries, Christians have opposed certain medical advances that are now common practice because of an interpretation of a Biblical passage which, today is viewed as an incorrect interpretation of such a passage.[20] Sickness was viewed as punishment from God, so treatment often involved tormenting the body to expel evil spirits. Priests were prohibited by papal decrees from shedding blood, thus preventing them from performing surgery. The church forbade dissections of human or animal cadavers, preventing any comprehensive study of anatomy or physiology. Other examples include being against inoculation and vaccination, and the use of

---

[19] "An Almost Absolute Value in History," [by] John T. Noonan, in: The Morality of Abortion: Legal and Historical Perspectives, edited, with an introduction, by John T. Noonan, Jr. (Cambridge, Mass.: Harvard University Press, c1970): pp. 1-59.

[20] A treatment of this subject can be found in *A History of the Warfare of Science with Theology in Christendom* by Andrew Dickson White LL.D.(Yale), L.H.D. (Columbia), PH.DR. (Jena) Late President and Professor of History at Cornell University Two Volumes Combined New York D. Appleton and Company 1898 Copyright, 1896By D. Appleton and Company.

anesthetics during childbirth. The use of anesthetics during childbirth is an excellent example.

Genesis 3:16: To the woman he said, "I will make your pains in childbearing very severe; with painful labor you will give birth to children. Your desire will be for your husband, and he will rule over you." New International Version.

A Scottish doctor, James Baker Simpson (1811-1870) heard of the pain and near fatal experience his mother went through in birthing him. He became chief obstetric assistant and later discovered that the use of chloroform would help women have less pain during childbirth. After 30 successful tests, he revealed his methods so that no woman would have to suffer through child birth. The reaction of the Scottish Calvinist Church was swift and furious. "What a Satanic invention!" they cried, "What a shame upon Edinburgh!" The Ecclesiastics objected to this rebellion against God, "Did not the Almighty pronounce this primal curse?" They asked. "Pain of childbirth was God's will. Now one of God's creatures, impiously rebelling against the divine command, had dared to frustrate God's will." In churches, the preachers warned pregnant women that, should they allow this devilish treatment to be administered on them, their children would be denied the sacrament of baptism. This warning must have worked, for we have record of Simpson complaining to a friend that "many of my lady patients had strong religious scruples against anesthesia. Most of them consult their ministers." The pulpit was not the only place from which the clergy attacked Simpson. They sent circulars to all doctors in Edinburgh which contained the following words:

To all seeming, Satan wishes to help suffering women but the upshot will be the collapse of society, for the fear of the

Lord which depends upon the petitions of the afflicted will be destroyed.[21]

The science has not changed, but over time we have come to believe that *interpretation* was wrong. Some examples were the Biblical basis against heart transplants and some Christian groups still opposed blood transfusions based on their interpretation of Biblical passages. One can believe in the inerrancy of scripture without believing in the inerrancy of human interpretation. What is argued is this: in light of new scientific discoveries, is it possible the interpretation of conception occurring at the time of fertilization is incorrect?

If it is acceptable to understand conception as occurring at individualization and not fertilization, it solves a number of practical problems such as the ethics of embryonic research, acceptability of various contraception methods that occur within the first two weeks of intercourse, the use of medical interventions in the cases of rape and incest, the use of in vitro fertilization (IVF) as an acceptable method for reproductive technology, and what to do with thousands of left over embryos in storage as a result of IVF, etc. It could also make regulation simpler in that registration could be required for embryonic stem cell or therapeutic cloning research and those who registered would be subject to surprise inspections where if research was taking place on embryos after the primitive streak was formed, they would be subject to sanction.

In summary, the church has not held one consistent view on the question of whether or not human life begins at the

---

[21] The Rejection of Pascal's Wager: A Skeptic's Guide to the Bible and the Historical Jesus Paul Tobin.

moment of fertilization. The church has been wrong in the past about advances in medical science, and there is a way to reconcile religion with the biological facts by understanding conception taking place at the time of the formulation of the primitive streak.

The next chapter looks at this topic in more depth, we also include in this book Dr. Michael West's testimony before Congress so that you can hear his arguments in context.

The Primitive Streak: Nature's Ethical Dividing Line for the Dilemma Concerning Therapeutic Cloning and Stem Cell Research

Cloning…Rarely does a scientific advance in the field of medicine promise such benefits while providing such ethical, moral and philosophical dilemmas. Similar questions were raised when the idea of in vitro fertilization (IVF) was first introduced nearly thirty years ago. However, society's worst fears of "test tube babies" were not realized, and thousands of people are now parents.

The importance of ongoing and widespread debate on the ethics of cloning, technically known as somatic nuclear transfer, and the debate on methodologies for the development of embryonic stem cells; including a discussion of limitations, guidelines, and the larger issue of the greatest good to society, is not only desirable but also essential.

Because "cloning" is itself a loaded term, thanks to pulp science fiction and sensational media coverage, it is necessary to define terms very closely – providing the accepted scientific definitions for its various forms. A distinction needs to be drawn between reproductive cloning (the use of the technology to produce an individual that is the exact genetic copy of another) and therapeutic cloning (which entails the creation of methodologies or therapies for the treatment of what are currently incurable diseases or conditions.)

This chapter is concerned solely with therapeutic cloning. Arguments against reproductive cloning – and the enormous number of deaths and deformities that occur in animal experiments – place this technology outside the realm of accepted morality. Most support an official ban on reproductive cloning because, while 99.9% of all scientists will not create a human clone on ethical grounds, there are a few groups of scientists who have vowed to create one as a way to advance the cause of reproductive medicine or religious goals.[22]

Therapeutic cloning and its companion technology of embryonic stem cell therapy (ES) should be supported by the political, social, and medical establishments of this country because of both the overwhelmingly positive benefits that can accrue from this research and the lack of compelling scientific or logically consistent ethical arguments against it. When these techniques are practiced within the guidelines proposed in this chapter, the technology can be leveraged to benefit those suffering from various diseases and injuries, while still safeguarding the sanctity of human life. Furthermore, therapeutic cloning does not need to be banned as a means of prohibiting reproductive cloning.

For many who support therapeutic cloning, the challenge is trying to define safeguards that would prevent reproductive cloning and protect the sanctity of human life. God has not left us on our own to solve this ethical dilemma. God has provided us a literal line that we can use as a definitive barrier we dare not cross. This barrier is called the primitive streak. The formation of the primitive streak provides a clear division between cellular life and human life, a dividing line, which, because it occurs in nature, can be agreed upon by biological researchers, is used currently in reproductive medicine, and could be ethically and popularly supported.

---

[22] Dr. Richard Seed, Brigitte Boisselier of the Clonaid/Raelians, Panos Zavos and Severino Antinori.

The fertilization of the egg cell or ovum by a sperm leads to a single cell called the "zygote". From this first cell, multiple rounds of cell division over the first week result in a microscopic ball of cells called the blastocyst or preimplantation embryo. Should the embryo implant in the uterus, the embryo will form what is called the primitive streak at approximately 14 days post-fertilization. The primitive streak is an elongated band of cells that forms along the axis of a developing fertilized egg early in gastrulation, its formation marks the start of gastrulation. The view of many bioethicists is the cellular life created before the formation of this line, the "primitive streak," is ethically and morally acceptable to use for research, including preimplantation embryos and, after it is formed, is not.

First, some biologists believe conception should be defined as when the primitive streak occurs and not at the point of fertilization. The primitive streak should mark the beginning of human development. Clearly there are those who believe once an embryo is formed by natural or artificial means, it must continue its development and become a human being. Anything short of that is murder. They believe once we "create life to kill it", we are one step closer to eugenics, Nazism, euthanasia, infanticide, and mercy killing. Some also believe labeling a human embryo as not human is the first step in designating some individuals as human and others as not, which has led in the past to slavery and discrimination. While some civilizations have offered human sacrifices for what were believed to have been noble purposes, none have been as barbaric as to create human life for the sole purpose of killing it, "clone and kill," which is what some argue is the very thing that is proposed with therapeutic cloning and stem cell research.

The problem with this argument is that it is simply wrong. It is not supported by science. Those who hold the position that human life begins at conception do not consistently apply it to reproductive practices, such as birth control and IVF. It takes an unfathomable leap in mental logic to understand how the copying of a cell, grown for 14 days in a petri dish, where stem cells could then be developed, would lead to the atrocities listed above. The truth of the matter is, while cloning and stem cell research is new, the ethics surrounding the issue are not. If we look at precedent, we as a society have already determined what is acceptable and what is not. What those who advocate for these new therapies are proposing does not break new ethical ground. It is simply just a different application.

Many bioethicists believe the use of cells prior to the formation of the primitive streak is ethically, morally and philosophically acceptable based on the arguments that follow.

Biological Arguments

First, borrowing an analogy from botany, a blastocyst is similar to a seed and, as distinct as a seed is from a plant, so are these seeds of cellular life different from human life. Similar to seeds, which must be planted before they can start the germination process, these blastocysts have no hope of becoming human life if they are not implanted in the womb. The discussion of when human life begins is often based on various milestones, such as the first heartbeat, the first brain wave, or the first breath. By restricting techniques to the seed stage rather than after the germination stage, one avoids the problem of trying to select the right relative milestone of human development. This restriction would also prevent the abortion of a human fetus for the purpose of extracting stem cells.

Second, it must be noted that 40-60%, possibly even as many as 80%, of all fertilized ova fail to attach to the uterus and are naturally destroyed as a result. The proposed solution of using these blastocyst cells is consistent with what naturally occurs. In other words, we are not placing a greater or lesser value on these seeds than God does.

While the correct term for the fertilized ovum that has not implanted itself in the uterus is *conceptus*, or preimplantation embryo, this small group of cells is not what most people would think of when using that term. From fertilization to the formulation of the primitive streak, the blastocyst is approximately 100-150 cells. This microscopic mass could fit within the period at the end of this sentence or on the point of a pen. Dr. Robin Cook feels it is unfortunate we use the term embryo when referring to this mass of cells, as the embryo does not form until week three. Embryonic stem cells do not come from embryos as the embryo as defined in biological terms has not been formed yet.

The third argument is that, at this point, there are no human body parts formed because the differentiated cells necessary to form body parts don't exist at this point either. In other words, what makes embryonic stems cells so powerful is they are undifferentiated, or nothing in particular, but have the *potential* to become anything such as a heart, finger or eye. ***Having the potential to become something is not the same as being something.*** It is hard to argue that this collection of cells constitutes a human being or even a human fetus.

The fourth reason that the formation of the primitive streak is instructive as a guideline is that prior to its formation no individuation has taken place. At approximately day 14, when the primitive streak forms, one can determine if the blastocyst cells will become one or two individuals or identical twins. "Developmental individuality, which is central to personhood, is not attained until the primitive body axis has begun to form."[23]

Reproductive Arguments

In the case of IVF, blastocysts or preimplantation embryos are grown for approximately 14 days before they are implanted in the uterus. IVF as it is commonly practiced results in an excess of fertilized ova being created. These embryos are often discarded once they are no longer needed for fertility purposes. If human life is created at the time of conception, then IVF would not be an accepted technology on ethical grounds, since it results in the creation of unwanted human embryos, which are destroyed or left in freezers never to reach their potential. Therefore, the use of embryos less than two weeks old is already the common, accepted practice in reproductive medicine and is viewed as ethically acceptable. If one were to ban therapeutic cloning or stem cell research because it entailed the destruction of blastocysts, then to be consistent, one would have to ban IVF as well.

Second, some common contraceptive practices[24] prevent the embryo from attaching to the uterus. This also takes place within the first two weeks of conception. What is proposed with the use of embryos not grown beyond 14 days or when the primitive streak is formed is consistent with these practices. If one were to ban therapeutic cloning or stem cell research because it entailed the destruction of nonimplanted embryos, then to be consistent, one would

---

[23] JAMA Dec. 27, 2000).
[24] RU-486, the "morning after pill" and IUDs.

have to ban certain types of contraceptive practices as well. In fact, to take the argument to absurd lengths, even common sexual reproduction would be prohibited because it results in the destruction of human embryos, those which don't attach to the uterus 40-80% of the time.

Third, current practice would indicate there is a property right involved. It is currently acceptable to donate organs, blood, retinas, and eggs. Courts are often asked to decide who "owns" human embryos in custody disputes. And the right to continue or terminate a pregnancy has been placed with the individual within some legal constraints. Embryonic stem cells are "totipotent," which means that they are potentially capable of forming any cell or tissue type needed in medicine. It would seem logical that one would be able to donate these cells, which have the potential to become organs, skin, brain cells, nerve receptors, or whatever else. Medicine should be free to use these cells as they would use skin for a skin graft or for an autogolous transplant. Individuals who feel the use of their DNA for this purpose is wrong could always choose not to donate their embryos or take advantage of these medical procedures.

The benefits of therapeutic cloning and stem cells are too important to be lost in the default Pro Life and Pro Choice arguments instead of the science that any embryologist knows to be true. The only logically consistent alternative position is the one advanced by the Roman Catholic Church, which opposes therapeutic cloning, embryonic stem cell research, birth control, in vitro fertilization, and abortion. However, the majority of Americans and many U.S. Roman Catholics don't support those views.

In addition, our society and our governmental agencies have already made many ethical decisions about the use of various types of contraception and IVF. Therapeutic cloning and the development of stem cells does not require science to do anything that has not already been deemed acceptable.

The Slippery Slope Argument

Some people may agree with the above arguments but believe a ban on therapeutic cloning is necessary as a means of preventing reproductive cloning. Most of the "slippery slope" arguments are based on the unfortunate history of Eugenics, avoiding Nazism or the creating of a "Brave New World," genetic engineering of the "perfect baby." Those who propose this argument never explain the steps necessary to go from copying a cell and using stem cells in necessary therapies to these socially unacceptable practices. And, of course, to embark upon that slippery slope, someone needs to break the law.

There are various arguments based on scientists being tempted to perform reproductive cloning. One is if a scientist were doing therapeutic cloning they might be tempted to do reproductive cloning given the work is done in private and perhaps in secret. If a scientist did place a cloned embryo in a uterus and the government caught it, would the government force the person to have an abortion? The government would face this and many other ethical problems. Currently the government does monitor and regulate work done in private labs and patient physician interactions which are also done in private. There is no reason why the government could not monitor and enforce the prohibition of reproductive cloning.

Another version of this argument is that the majority of scientists are against it because the failure rate is so high, 95-97%. But what if over time, the success rate became

99%? Would that not change a scientist's position and perhaps tempt them to try reproductive cloning? The problem with this argument is that in order to prevent the advancement of cloning knowledge, *all* cloning research would have to be prevented, especially in animals, thus severely limiting science and medicine.

At present, reproductive cloning is legal and there are perhaps only a half dozen people in the *world* who have an interest in reproductive cloning. It is hard to imagine once therapeutic cloning became *illegal*, more scientists and physicians would want to pursue reproductive cloning, especially since it is almost universally condemned. What would be the incentive for those to break the law? Those that did would be potentially risking losing their lab or medical practice and would probably face malpractice or other civil law suits. Their reputation would be ruined. If a scientist were willing to risk all of this and, in addition, suffer criminal and civil punishment, then an additional ban on therapeutic cloning would not stop him or her. However, it would stop all of the legitimate research that could be done to help society. Banning therapeutic cloning research will not deter those who are determined to experiment with reproductive cloning. Creating a climate where research is encouraged and monitored is a better way to prevent the erosion of ethical research.

As can be seen from the above arguments, permitting therapeutic cloning is not the first step toward reproductive cloning. Research and treatment using therapeutic cloning and stem cells are legitimate areas of research. Banning therapeutic cloning because of what it could lead to would be like banning clinical trials in medicine because it could lead to atrocities as experienced in Germany when they conducted human experiments. It is hard to imagine how much worse off we would be as a society if clinical trials had never been allowed. And the same will be true if we can't conduct research into these important areas.

Another argument is this: why not just use adult stems cells where there is no argument and avoid the issue of embryonic stems cells completely? Some even argue that adult stems cells are better from a scientific perspective than embryonic stem cells. And those who admit that adult stem cells do not have the same advantages as embryonic stem cells argue that with more money and research, they will be. The response to this argument is it avoids the discussion of using embryonic stems cells as opposed to refuting it on scientific or ethical grounds. They still need to refute the arguments as to why the use of embryonic stem cells is not acceptable. There is no reason a both/and approach could not be taken.

Proposed Solution

Many would propose it would be illegal to grow human embryos beyond 14 days or when the primitive streak is formed, for any purpose, whether for stem cell research or therapeutic cloning, whether for experimentation or commercialization, and that it would be illegal to implant a cloned human embryo in a womb.

The formulation of the primitive streak, given to us by nature, is a stop sign which we do not pass. If we heed its warning, we are in no danger of falling down the slippery slope. This proposed solution provides for the advancement of science without sacrificing the sanctity of human life. It is based on well-understood embryological science and is consistent with how we practice reproductive medicine today. It not only solves the problem for therapeutic cloning and stem cell research, but it also provides an acceptable solution of what to do with excess embryos in fertility clinics, estimated to be between 30,000 and 50,000. It provides an ethical solution consistent with what is morally acceptable in the area of reproductive medicine. It is consistent with previous legislation and regulations, e.g., FDA guidelines that have been passed in this area. This solution is also consistent with what the United Kingdom has legislated and, as such, could be the basis for global standards in the European Union, United Nations and the World Health Organization. As such the primitive streak should be adopted as the guiding principle in solving these ethical questions and the formation of legislative guidelines.

Dictionary Definitions of Conception

Norman E. Andersen at the time, reference librarian at Gordon-Conwell Theological Seminary provided the following research on the definition of conception. This insight is critical as the basis for objecting to the use of embryonic stem cell research is based on the definition that conception occurs at the time of fertilization.

You may notice a lot of imprecision in definitions. For example, the definition of "conception" according to The American Heritage Illustrated Encyclopedic Dictionary (1987) – a dictionary that is usually pretty good, but failed us this time:

"The fertilization of an egg cell by a sperm in the uterus to form an embryo capable of survival and maturation in normal conditions."

Actually, fertilization normally takes place in a fallopian tube. The immediate result is a zygote, which divides repeatedly and thus becoming a ball of cells called a morula. The morula develops a cavity, thereby becoming a blastula (so far following the chart in the AMA Encycl. of Med. on p. 446). A modified stage of the blastula is the blastocyst. The embryoblast is the inner cell mass at the embryonic pole of the blastocyst concerned with the formation of the body of the embryo per se (Stedman's).

The misimpression would be that a zygote is not an embryo in some definitions of "embryo." However, there is a process of cellular differentiation, such that not all of the cells go into what becomes the embryo per se (to use Stedman's term), which in turn becomes the fetus. It is interesting to compare definitions of "embryo" with definitions of "conceptus."

Stedman's:

Embryo: In man, the developing organism from conception until approximately the end of the second month; developmental stages from this time to birth are commonly designated as fetal."

Conceptus: Pl. concepti ... The product of conception, i.e. embryo and membranes.

The Language of Sex from A-Z:

Conceptus: The products of conception, including the embryo or fetus, the fetal membranes, the amniotic fluid, and the fetal portion of the placenta.

Embryo: The developing organism between the stage of a fertilized ovum and the fetal stage. In humans, an organism is considered an embryo from approximately the 14th day of gestation until the 55th day...

The Complete Dictionary of Sexology:

Conceptus: In the strict sense, the embryonic blastocyst when it implants in the uterine mucosa. A less specific usage refers to the zygotic product of conception, the embryo, or even the fetus.

Embryo; Any organism in its earliest stages of development when the major organ systems and the main external features are established. In humans, the embryonic stage extends from the second to the end of the eighth week of gestation...

Taber's

Conceptus: The products of conception.

Embryo: In humans, stage of development between the 2nd and 8th weeks inclusive.

I can't help definitional inconsistencies! I notice that Taber's uses even the term "zygote" in more than one way, so that it can refer either to the fertilized ovum (s.v. zygote) or to the stage of development from the fertilized ovum right through to implantation in the uterus (s.v. embryo).

-- Norman E. Anderson

Some points in the history of the idea:

Exodus 21:22-23 (LXX): "If two men fight and strike a woman with child and she miscarry of an embryo, atonement shall be made by a fine. According as the husband of the woman shall with a judicial decision lay upon him, he shall pay: but if the child be completely organized he shall give, life for life..."

--> The Septuagint Bible: The Oldest Version of the Old Testament, in the translation of Charles Thomson; as edited, revised and enlarged by C. A. Muses (Indian Hills, Colorado: Falcon's Wing Press, 1954). Compare Psalm 139:13-16 and Jeremiah 1:4-5.

Apostolic Constitution, Book 7, Section 1, §3: "Thou shalt not slay thy child by causing abortion, nor kill that which is begotten; for 'everything that is shaped, and has received a soul from God, if it be slain, shall be avenged as being unjustly destroyed' [Exodus 21:23, LXX]."

--> Constitutions of the Holy Apostles, edited, with notes, by James Donaldson, in the Ante-Nicene Fathers: Translations of the Writings of the Fathers down to A.D. 325, Alexander Roberts and James Donaldson, editors ; revised by A. Cleveland Coxe; v. 7 (1975 printing): p. 466.

The Prophetic Scriptures 48: "Peter says in the Apocalypse, that abortive infants shall share the better fate ..."

The Prophetic Scriptures 50: "the embryo is a living thing; for that the soul entering into the womb after it had been by cleansing prepared for conception, and introduced by one of the angels who preside over generation, and who know the time for conception, moves the woman to intercourse; and that, on the seed being deposited, the spirit, which is in the seed, is, so to speak, appropriated, and is thus assumed into conjunction in the process of formation.... And the barren are barren for this reason, that the soul, which unites for the deposit of the seed, is not introduced so as to secure conception and generation."

--> As quoted in The Encyclopedia of Christianity, v. 1 (1964): pp. 20-21; from: Liturgies and Other Documents of the Ante-Nicene Period (Edinburg, 1872)

The following quotations and citations are from Noonan (1970), which is fully cited below. I haven't checked all of his references.

"Augustine, commenting on a Latin translation from the Septuagint, observed that at Exod. 21 the question of ensoulment was usually raised, and 'because the great question about the soul is not to be hastily decided by unargued and rash judgment, the law does not provide that the act pertains to homicide, for there cannot yet be said to be a live soul in a body that lacks sensation when it is not formed in flesh and so not yet endowed with sense.'* This was a distinction accepted out of a cautious agnosticism on ensoulment; both Jerome and Augustine affirmed that, in fact, man did not know when the rational soul was given by God.**

"As far as Jerome and Augustine were concerned, the theoretical distinction led to no difference in moral disapprobation. They simply adopted language broad enough to condemn both contraceptive acts and acts destroying the fetus after conception.***" (p. 15)

* Augustine, On Exodus 21.80, CSEL 28(2).147.
** Augustine, De Origine Animae 4.4 (PL 44.527); Jerome, On Ecclesiastes 2.5.

*** Jerome, Epistle 22, To Eustochium 13, CSEL 54.160-61; Augustine, De Nuptiis et Concupiscentia 1.15.17, CSEL 42.229-30.

Thomas Aquinas "was clear that there was actual homicide when an ensouled embryo was killed.* He was equally clear that ensoulment did not take place at conception.**" (p. 23)

* Thomas Aquinas, Summa Theologica 2.2.64.8, reply to objection 2.
** Thomas Aquinas, In Libros IV Sententiarum 3.1.1.

"Medico-Legal Questions, by a Roman physician, Paolo Zacchia [was published in 1621]. In his learned treatise on medical aspects of the canon and civil laws Zacchia attacked the prevailing interpretation of Aristotle which envisioned the fetus progressing by stages from vegetable ensoulment to animal ensoulment to rational ensoulment. This 'metamorphosis of souls,' he declared, was 'an imaginary thing.' Belief that the rational soul was in fact instilled after forty days rested on no evidence that the rational soul was then in operation; nor could the movement of the fetus have any significance in showing the presence of a rational soul. Those who argued that there was a rational soul at some time in the embryo, but at some time after conception, were thus entangled in 'absurdities' in trying to show the basis of their conviction. On the contrary, a true Thomistic view of the unity of man required that there be a single human soul from the beginning of the existence of a new fetus. The rational soul, Zacchia argued, must be 'infused in the first moment of conception.'"* (pp. 34-35)

However, "The theory of Zacchia had no immediate impact on the theologians dealing with abortion... By the eighteenth century Constantino Roncaglia of the Congregation of the Mother of God contended in analyzing the sin of abortion that it was 'most probable' that the fetus was ensouled at the instant of conception or 'at least from the third or seventh day.'**" (pp. 35-36)

*Paolo Zacchia, Quaestiones Medico-Legales (Lyons, 1701): 9.1; 9.5.
** Constantino Roncaglia, Universale Moralis Theologia ad Usum Confessionarum (Lucca, 1834): 11.1.2.3.

Regarding Thomas Fienus (De Feynes), who argued in a treatise published in 1620 that "the rational soul is infused on the third day" after conception, and Paolo Zacchia, The Encyclopedia of Bioethics (1978) comments:

"The opinion of Fienus and Zacchia on immediate animation, although not rejected by the Church, did run into considerable opposition from theologians. It was objected to on three specific grounds: It was contrary to the Scriptures, to the universal opinion of theologians, and to the practice of the Church. It might be surprising that an opinion that had so much authority against it could survive. But the fact is that it did take hold and gradually replaced theories of delayed animation." (v. 1, p. 12, col. [1])

Vatican II: "For God, the Lord of life, has conferred on men the surpassing ministry of safeguarding life -- a ministry which must be fulfilled in a manner which is worthy of man. Therefore from the moment of its conception life must be guarded with the greatest care, while abortion and infanticide are unspeakable crimes."

--> Pastoral Constitution on the Church in the Modern World (1965): part 2, chapter 1, §51; as found in: Gonzalez (1967?): pp. 566-567.

According to Noonan (1970), the Vatican II "declaration was the first statement ever made by a general council of the Church on abortion." (p. 46)

According to The Encyclopedia of Bioethics (1978): "The Church has made no positive teaching statement regarding the time of infusion of the human soul, and this is true of Vatican II as well as of earlier documents... The only opinion the Church has condemned is that of Ioannes Marcus, that the human soul is not infused until birth." (v. 1, p. 12)

The above pinpoints some of the steps in the Roman Catholic development of the idea that "life begins at conception." The Protestant reformers took a somewhat different tack, one which implied the same conclusion. To quote again from the Encyclopedia of Bioethics (1978):

"The major reformers -- Martin Luther (1483-1546), Philip Melanchthon (1497-1560), and John Calvin (1509-1564) -- were at least as conservative as their Roman Catholic counterparts on the issues of ensoulment and the gravity of abortion. Indeed, some historians ... [George Huntston Williams is mentioned] believe that they indirectly but significantly contributed to the present papal position on the subject.

"The reformers insisted upon the full humanity of the fetus from the time of conception. Their insistence arose, however, less from attention to the abortion issue itself than from their concern about the doctrines of original sin and predestination. Full humanity of the conceptus was believed necessary if the mind and spirit as well as the body of nascent life were to be involved in the consequences of the human Fall." (v. 1, p. 14, col. [1])

The following three quotations are from Montgomery (1969):

"'Creationism,' or (better) 'concreationism,' is a theological position held by Pelagius, Peter Lombard, St. Thomas,* the Roman Catholic ordinary magisterium ..., and by most Calvinists.** This view affirms that God creates souls ex nihilo and supplies them to developing individuals at conception or during the intrauterine period." (p. 76)

"'Materialistic' traducianism holds either that parents generate from inanimate matter not only the body but also the soul of the child, or that the soul is actually contained in the sperm*** and conveyed by organic generation. More attractive by far has been 'spiritual' traducianism, often called 'generationism,' which asserts that the soul of the child derives from the souls of the parents. Augustine**** in opposing the Pelagians and in his insistence on man's total depravity, held to generationism, as did Luther and most theologians influenced by him." (pp. 78-79)

"For the traducianist, it would be absurd to regard the individual as commencing later than conception, for even his soul derives from his parents. For most creationists, the moment of conception is the point when the soul is bestowed." (p. 79)

* Thomas Aquinas held a mediate animation theory, asserting that the male receives his 'rational soul' forty days after conception and the female eighty to ninety days. See his Summa Theologica, Part I, question 75, article i; cf. question 76, article iii ad 3; question 118, article ii ad 2 (so cited by Montgomery on p. [69]).

** The Encyclopedia of Bioethics (1978) says that Melancthon was a creationist too. See v. 1, p. 14, col. [1].

*** The role of the ovum was not understood until 1875, having only been discovered in 1827 by Karl Ernest von Baer. See Noonan (1970): p. 38.

**** Augustine, Epistles 166.8.25-26; 190.4.14-15 (so cited by Montgomery).

The Encyclopedia of Bioethics (1978) discusses a number of proposed answers to the question, When does truly human life begin? For example, it describes this nuanced creationist theory, in which mere conception is inadequate for full human ensoulment:

"Joseph Donceel has based a delayed hominization or delayed animation theory on the Thomistic concept of hylomorphism (viz., there is a complementarity between the material and formal aspects of being.). The form, which in this case is the soul, is received only into matter capable of receiving it... Donceel concludes that the least that must be present before admitting a human soul is the availability of these organs: senses, the nervous system, the brain, and especially the cortex." (v. 1, p. 18)

Reference is given to: "Immediate Animation and Delayed Hominization," by Joseph F. Donceel, Theological Studies; 31 (1970): pp. 76-105.

Above I quoted from the Encyclopedia of Bioethics, Warren T. Reich, editor in chief (New York: Free Press, 1978), which has a useful historical survey, dissection of the abortion issue, and summary of many of the positions. A more recent edition came out in 1995. The recast abortion article from it appears also in: The Ethics of Sex

and Genetics: Selections from the Five-Volume Macmillan
Encyclopedia of Bioethics, Revised Edition, Warren
Thomas Reich, editor in chief (New York: Macmillan
Reference USA, c1998; "Macmillan Information Now
Encyclopedia")

For another useful survey of the issue, let me refer you to
the following sources:

Encyclopedia of Ethics, Lawrence C. Baker, editor;
Charlotte B. Becker, associate editor (New York: Garland,
1992)

Given an interest of the Evangelical perspective, you may
profit as well from the article on abortion in:

Evangelical Dictionary of Theology, edited by Walter A.
Elwell (Grand Rapids, Mich.: Baker Book House, c1984)

Here are full citations of items mentioned in short form
above (with Jewett thrown in because of relevance):

Gonzalez (1967) Pastoral Constitution on the Church in the
Modern World, promulgated by His Holiness, Pope Paul VI
on December 7, 1965; as found in: The Sixteen Documents
of Vatican II ... compiled by J. L. Gonzalez and the
Daughters of St. Paul (Boston, Mass.: Daughters of St.
Paul, [1967?]): pp. [513]-624. Abortion is mentioned also
on p. 539 = Part 1, Chapter 2, §27.

Jewett (1969). "The Relationship of the Soul to the Fetus,"
by Paul K. Jewett, in: Birth Control and the Christian: A
Protestant Symposium on the Control of Human
Reproduction, edited by Walter O. Spitzer and Carlyle L.
Saylor (Wheaton, Ill.: Tyndale House Publishers, c1969):
pp. [49]-66.

Montgomery (1969). "The Christian View of the Fetus," by John Warwick Montgomery, in: Spitzer and Saylor (1969): pp. [67]-89.

Noonan (1970) "An Almost Absolute Value in History," [by] John T. Noonan, in: The Morality of Abortion: Legal and Historical Perspectives, edited, with an introduction, by John T. Noonan, Jr. (Cambridge, Mass.: Harvard University Press, c1970): pp. 1-59.

Bibliography:

John Riddle *in* Contraception and Abortion from the Ancient World to the Renaissance (Harvard University Press, 1992)

Abortion: Ancient Christian Beliefs, ReligiousTolerance.org

A History of Contraception from Antiquity to the Present Day Angus McLauren, Blackwell 1990

The Immortal Cell: One Scientist's Quest to Solve the Mystery of Human Aging [Hardcover] by Michael West

When Did I Begin?: Conception of the Human Individual in History, Philosophy and Science (Conception of the Human Individual in History and Philosophy) [Paperback] by Norman Ford, Mary Warnock

The Prenatal Person: Ethics from Conception to Birth Paperback by Norman M. Ford

Babies by Design: The Ethics of Genetic Choice
[Paperback]
by <u>Ronald M. Green</u>

<u>http://religion.dartmouth.edu/people/ronald-michael-green</u>

I include here a selection of Dr. Michael West's Congressional Testimony. A bibliography of his Congressional testimony follows that is hyperlinked so you can read it in context. You can use the references to get a copy of the hearings from the Government Printing Office.

Congressional Testimony of Dr. Michael West

DATE: July 18, 2001

Mr. Chairman and members of the Subcommittee on Labor, Health and Human Services, Education, and Related Agencies, my name is Michael D. West and I am the President and Chief Executive Officer of Advanced Cell Technology, Inc., a biotechnology company based in Worcester, Massachusetts. A copy of my curriculum vitae is presented in Appendix A.

INTRODUCTION

I am pleased to testify today in regard to the new opportunities and challenges associated with human embryonic stem (ES) cell and nuclear transfer (NT) technologies. I will begin by describing the bright promise of these twin and interrelated technologies and then attempt to correct some misunderstandings relating to their application in medicine.

It may be useful to point out that I think of myself as pro-life in that I have an enormous respect for the value of the individual human life. Indeed, in my years following college I joined others in the protest of abortion clinics. My goal was not to send a message to women that they did not have the right to choose. My intent was simply to urge them to reconsider the destruction of a developing human being. Despite my strong convictions about the value of the individual human life, in 1995 I organized the collaboration between Geron Corporation and the laboratories of Dr. James Thomson and John Gearhart to isolate human embryonic stem cells and human embryonic germ cells from human embryos and fetuses respectively. My reasons were simple. These technologies are entirely designed to be used in medicine to alleviate human suffering and to save human life. They are, in fact, pro-life. The opponents that argue they destroy human lives are simply and tragically mistaken. Let me explain why this is the case.

Human ES Cells

We are composed of trillions of individual living cells, glued together like the bricks of a building to construct the organs and tissues of our body. The cells in our bodies are called "somatic cells" to distinguish them from the "germ line", that is, the reproductive cells that connect the generations. We now know that life evolved from such single-celled organisms that dominated all life some one billion years ago.

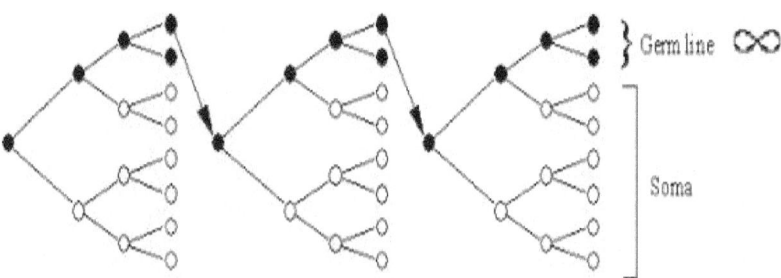

Figure 1. **The Distinction Between the Life of Cells and the Life of an Individual Human Being in the Human Life Cycle. Biological, that is, "cellular life" began with the origins of life on earth. The beginning of the life of an individual human being is linked to the appearance of somatic cells, that is, cells committed to form the human body.**

Therefore, in answer to the question of when life begins, we must make a crucial distinction. Biological life, that is to say, "cellular life" has no recent beginnings. Our cells are, in fact, the descendants of cells that trace their beginnings to the origin of life on earth. However, when we speak of an individual human life, we are speaking of the communal life of a multicellular organism springing from the reproductive lineage of cells. The individual human life is a body composed of cells committed to somatic cell lineages. All somatic cells are related in that they originate from an original cell formed from the union of a sperm and egg cell.

The fertilization of the egg cell by a sperm leads to a single cell called the "zygote". From this first cell, multiple rounds of cell division over the first week result in a microscopic ball of cells with very unusual properties. This early embryo, called the "preimplantation embryo", has not implanted in the uterus to begin a pregnancy. It is estimated that approximately 40% of preimplantation embryos formed following normal human sexual reproduction fail to attach to the uterus and are naturally destroyed as a result.

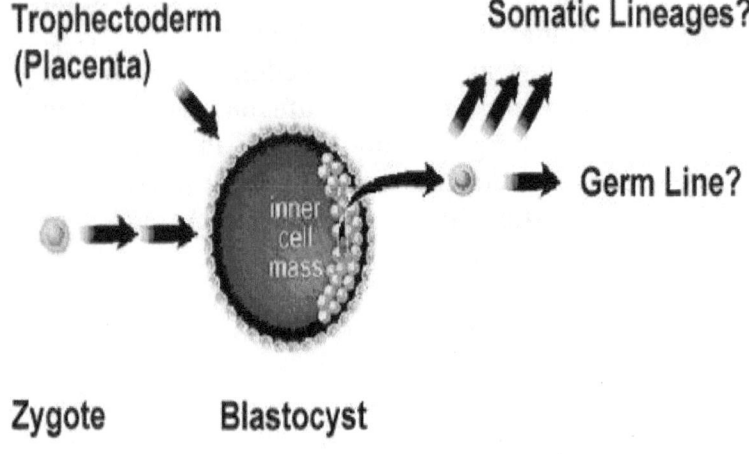

Trophectoderm
(Placenta)

Somatic Lineages?

Germ Line?

inner
cell
mass

Zygote   Blastocyst

Figure 2. The Blastocyst Stage of the Preimplantation Embryo. At the blastocyst stage of the preimplantation embryo, the external cells called the "trophectoderm" are destined to attach to the uterus and form the placenta. The remaining cells, the Inner Cell Mass (ICM) are completely undifferentiated and have not committed to any somatic cell lineage vs. the germ line.

From the above it should be clear that at the blastocyst stage of the preimplantation embryo, no body cells of any type have formed, and even more significantly, there is strong evidence that not even the earliest of events in the chain of events in somatic differentiation have been initiated. A simple way of demonstrating this is by observing subsequent events.

Should the embryo implant in the uterus, the embryo, at approximately 14 days post fertilization will form what is called the primitive streak, this is the first definition that these "seed" cells will form an individual human being as opposed to the forming of two primitive streaks leading to identical twins. Rarely two primitive streaks form that are not completely separated leading to conjoined or Siamese twins. In addition, rarely, two separately fertilized egg cells fuse together to form a single embryo with two different cell types. This natural event leads to a tetragammetic chimera, that is a single human individual with some of the cells in their body being male from the original male embryo, and some cells being female from the original female embryo. These and other simple

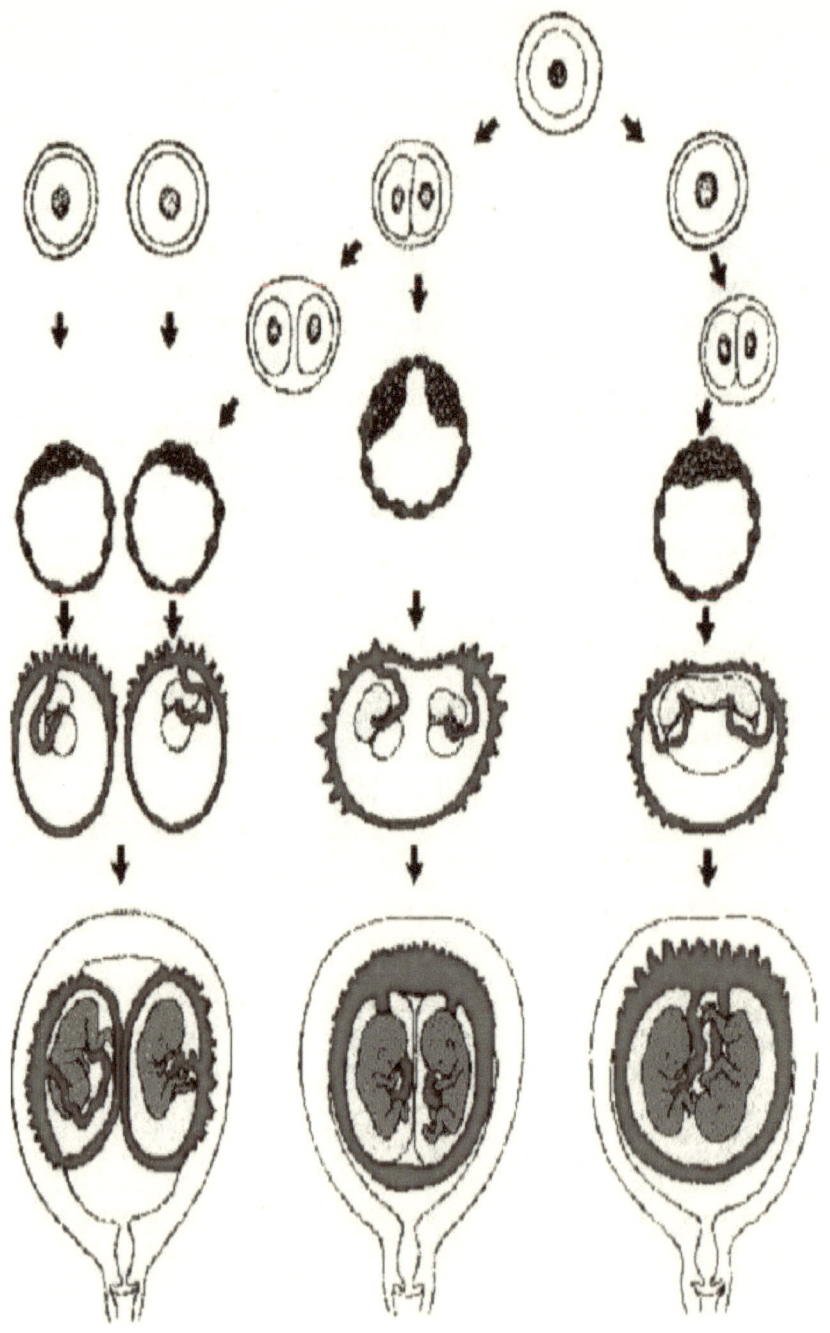

Figure 3.   The lack of Individuation of The Blastocyst Stage Embryo. Lessons from nature indicate that the blastocyst-stage preimplantation embryo has not individualized. On the left fraternal or nonidentical twins form from independently-fertilized egg cells. Identical twins form from a single ICM breaking into two ICMS (center diagram) or by two primitive streaks forming on one ICM (right diagram).

lessons in embryology teach us that despite the dogmatic assertions of some theologians, the evidence is decisive in support of the position that an individual human life, as opposed to merely cellular life, begins with the primitive streak, (i.e. after 14 days of development). Those who argue that the preimplantation embryo is a person are left with the logical absurdity of ascribing to the blastocyst personhood when we know, scientifically speaking, that no individual exists (i.e. the blastocyst may still form identical twins).

Human ES cells are nothing other than ICM cells grown in the laboratory dish. Because these are pure stem cells uncommitted to any body cell lineage, they may greatly improve the availability of diverse cell types urgently needed in medicine. Human ES cells are unique in that they stand near the base of the developmental tree. These cells are frequently designated "totipotent" stem cells, meaning that they are potentially capable of forming any cell or tissue type needed in medicine. These differ from adult stem cells that are "pluripotent" that is, capable of forming several, but only a limited number, of cell types. An example of pluripotent adult stem cells are the bone marrow stem cells now widely used in the treatment of cancer and other life-threatening diseases.

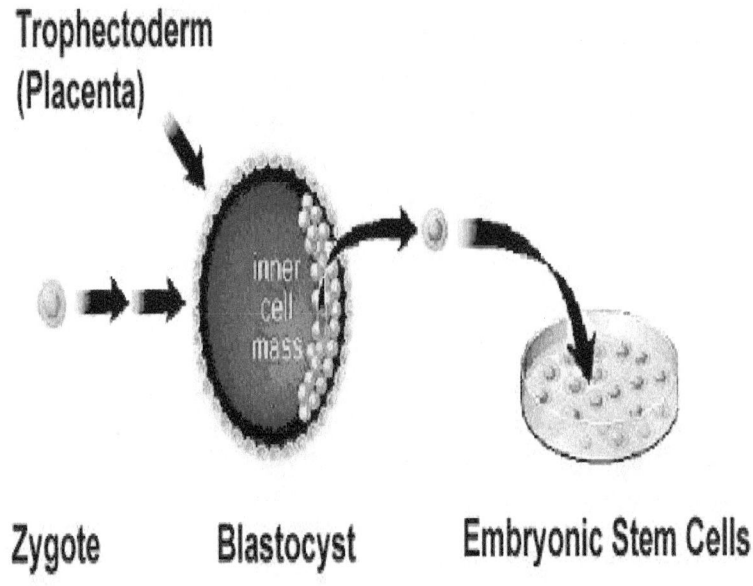

Trophectoderm (Placenta)

Zygote   Blastocyst   Embryonic Stem Cells

Figure 4. Human Embryonic Stem (ES) Cells are Inner Cell Mass (ICM) Cells Cultured in the Laboratory Dish. Human ES cells were first cultured for long periods of time in the laboratory by Dr. James Thomson of the University of Wisconsin at Madison.

Some have voiced objection to the use of human ES cells in medicine owing to the source of the cells. Whereas the use of these new technologies has already been carefully debated and approved in the United Kingdom, the United States lags disgracefully behind. I would like to think it is our goodness and our kindness as a people that generates our country's anxieties over these new technologies. Indeed, early in my life I might have argued that since we don't know when a human life begins, it is best not to tamper with the early embryo. That is to say, it is better to be safe than sorry. I believe many U.S. citizens share this initial reaction. But, with time the facts of human embryology and cell biology will be more widely understood. As the Apostle Paul said: "When I was a child, I spoke as a child, I understood as a child, I though as a child: but when I became a man, I put away childish things." (I Cor 13:11) In the same way it is absolutely a matter of life and death that policy makers in the United States carefully study the facts of human embryology and stem cells. A child's understanding of human reproduction simply will not suffice and such ignorance could lead to disastrous consequences.

With appropriate funding of research, we may soon learn to direct these cells to become vehicles of lifesaving potential. We may, for instance, become able to produce neurons for the treatment of Parkinson's disease and spinal cord injury, heart muscle cells for heart failure, cartilage for arthritis and many others as well. This research has great

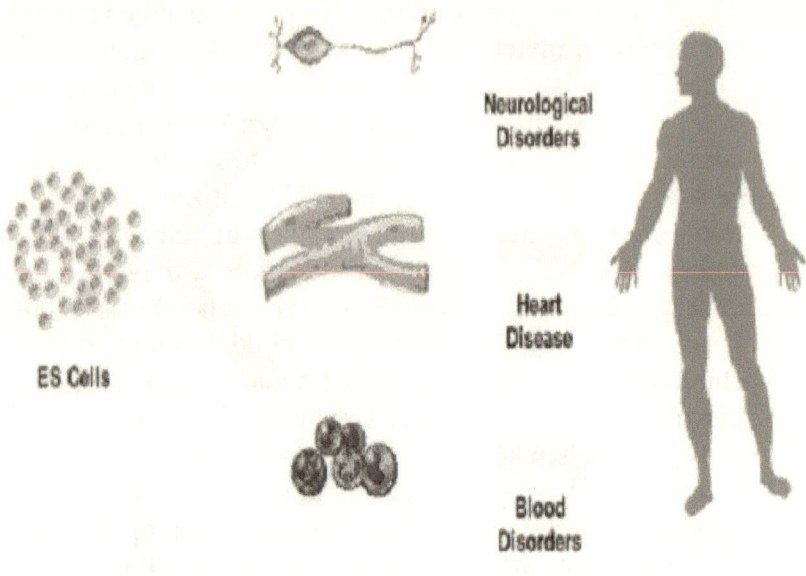

Figure 5. Using Embryonic Stem (ES) cells in human therapy. Human ES cells are immortal while cultured in the undifferentiated state and can theoretically lead to any cell or tissue type in the human body.

potential to help solve the first problem of tissue availability, but the technologies to direct these cells to become various cell types in adequate quantities remains to be elucidated. Because literally hundreds of cell types are needed, thousands of academic research projects need to be funded, far exceeding the resources of the biotechnology industry.

As promising as ES cell technology may seem, it does not solve the remaining problem of histocompatibility. Human ES cells obtained from embryos derived during in vitro fertilization procedures, or from fetal sources, are essentially cells from another individual (allogeneic). Several approaches can be envisioned to solve the problem of histocompatibility. One approach would be to make vast numbers of human ES cell lines that could be stored in a frozen state. This "library" of cells would then offer varied surface antigens, such that the patient's physician could search through the library for cells that are as close as possible to the patient. But these would likely still require simultaneous immunosuppression that is not always effective. In addition, immunosuppresive therapy carries with it increased cost, and the risk of complications including malignancy and even death.

Another theoretical solution would be to genetically modify the cultured ES cells to make them "universal donor" cells. That is, the cells would have genes added or genes removed that would "mask" the foreign nature of the cells, allowing the patient's immune system to see the cells as self. While such technologies may be developed in the future, it is also possible that these technologies may carry with them unacceptably high risks of rejection or other complications that would limit their practical utility in clinical practice.

Given the seriousness of the current shortage of transplantable cells and tissues, the FDA has demonstrated a willingness to consider a broad array of options including the sourcing of cells and indeed whole organs from animals (xenografts) although these sources also pose unique problems of histocompatibility. These animal cells do have the advantage that they have the potential to be genetically engineered to approach the status of "universal donor" cells, through genetic engineering. However as described above, no simple procedure to confer such universal donor status is known. Most such procedures are still experimental and would likely continue to require the use of drugs to hold off rejection, drugs that add to health care costs, and carry the risk of life-threatening complications.

Therapeutic Cloning

An extremely promising solution to this remaining problem of histocompatibility would be to create human ES cells genetically identical to the patient. While no ES cells are known to exist in a developed human being and are therefore not available as a source for therapy, such cells could possibly be obtained through the procedure of somatic cell nuclear transfer (NT), otherwise known as cloning technology. In this procedure, body cells from a patient would be fused with an egg cell that has had its nuclear DNA removed. This would theoretically allow the production of a blastocyst-staged embryo genetically identical to the patient that could, in turn, lead to the production of ES cells identical to the patient. In addition, published data suggests that the procedure of NT can "rejuvenate" an aged cell restoring the proliferative capacity inherent in cells at the beginning of life. This could lead to cellular therapies with an unprecedented opportunity to improve the quality of life for an aging population.

The use of somatic cell nuclear transfer for the purposes of dedifferentiating a patient's cells and obtain autologous undifferentiated stem cells has been designated "Therapeutic Cloning" or alternatively, "Cell Replacement by Nuclear Transfer". This terminology is used to differentiate this clinical indication from the use of NT for the cloning of a child that in turn is designated "Reproductive Cloning". In the United Kingdom, the use of NT for therapeutic cloning has been carefully studied by their Embryology Authority and formally approved by the Parliament.

Ethical Considerations

Ethical debates often center over two separate lines of reasoning. Deontological debates are, by nature, focused on our duty to God or our fellow human being. Teleological arguments focus on the question of whether the ends justify the means. Most scholars agree that human ES cell technology and therapeutic cloning offer great pragmatic merit, that is, the teleological arguments in favor of ES and NT technologies are quite strong. The lack of agreement, instead, centers on the deontological arguments relating to the rights of the blastocyst embryo and our duty to protect the individual human life.

I would argue that the lack of consensus is driven by a lack of widespread knowledge of the facts regarding the origins of human life on a cellular level and human life on a somatic and individual level. So the question of when does life begin, is better phrased "when does an individual human life begin." Some dogmatic individuals claim with the same certainty the Church opposed Galileo's claim that the earth is not the center of the universe, that an individual human life begins with the fertilization of the egg cell by the sperm cell. This is superstition, not science. The belief that an individual human being begins with the fertilization of the egg cell by the sperm cell is without basis in scientific fact or, for that matter, without basis in religious tradition.

All strategies to source human cells for the purposes of transplantation have their own unique ethical problems. Because developing embryonic and fetal cells and tissues are "young" and are still in the process of forming mature tissues, there has been considerable interest in obtaining these tissues for use in human medicine. However, the use of aborted embryo or fetal tissue raises numerous issues ranging from concerns over increasing the frequency of elected abortion to simple issues of maintaining quality controls standards in this hypothetical industry. Similarly, obtaining cells and tissues from living donors or cadavers is also not without ethical issues. For instance, an important question is, "Is it morally acceptable to keep "deceased" individuals on life support for long periods of time in order to harvest organs as they are needed?"

The implementation of ES-based technologies could address some of the ethical problems described above. First, it is important to note that the production of large numbers of human ES cells would not in itself cause these same concerns in accessing human embryonic or fetal tissue, since the resulting cells have the potential to be grown for very long periods of time. Using only a limited number of human embryos not used during in vitro fertilization procedures, could supply many millions of patients if the problem of histocompatibility could be resolved. Second, in the case of NT procedures, the patient may be at lower risk of complications in transplant rejection. Third, the only human cells used would be from the patient. Theoretically, the need to access tissue from other human beings could be reduced.

Having knowledge of a means to dramatically improve the delivery of health care places a heavy burden on the shoulders of those who would actively impede ES and NT technology. The emphasis on the moral error of sin by omission is widely reflected in Western tradition traceable to Biblical tradition. In Matthew chapter 25 we are told of the parable of the master who leaves talents of gold with his servants. On servant, for fear of making a mistake with what was given him, buries the talent in the ground. This servant, labeled "wicked and slothful" in the Bible, reminds us, that simple inaction, when we have been given a valuable asset, is not just a lack of doing good, but is in reality evil. There are times that it is not better to be safe than sorry.

Historically, the United States has a proud history of leading the free world in the bold exploration of new technologies. We did not hesitate to apply our best minds in an effort to allow a man to touch the moon. We were not paralyzed by the fear that like the tower of Babel, we were reaching for the heavens. But a far greater challenge stands before us. We have been given two talents of gold. The first, the human embryonic stem cell, the second, nuclear transfer technology. Shall we, like the good steward, take these gifts to mankind and courageously use them to the best of our abilities to alleviate the suffering of our fellow human being, our will we fail most miserably and bury these gifts in the earth? This truly is a matter of life and death. I urge you to stand courageously in favor of existing human life. The alternative is to inherit the wind.

Senate Appropriations Labor, Health and Human Services, and Education Subcommittee Stem Cell Hearing

Testimony of Michael D. West, Ph.D.

August 1, 2001

Testimony of Michael D. West, Ph.D. President & CEO Advanced Cell Technology, Inc., to the Senate Appropriations Labor, Health and Human Services, and Education Subcommittee Hearing to discuss the ethical and intellectual property rights issues involved in stem cell research.

Mr. Chairman and members of the Subcommittee on Labor, Health and Human Services, Education, and Related Agencies, my name is Michael D. West and I am the

President and Chief Executive Officer of Advanced Cell Technology, Inc., a biotechnology company based in Worcester, Massachusetts. A copy of my curriculum vitae is presented in Appendix A.

INTRODUCTION

I am pleased to testify today regarding to the emerging medical applications of human Embryonic Stem (ES) cell and Nuclear Transfer (NT) technology. These advances have unusually broad fields of application in medicine and are worthy of debate in the light of science, law, and ethics.

The Science

Human ES cells are profoundly unique in the history of cell biology. By way of perspective, all normal human cells grown in the laboratory dish over the last 40 years show evidence of aging upon long-term culture in the laboratory (a property called cellular mortality). Therefore, skin cells, bone cells, even adult stem cells will divide a mere 60-100 doublings after being placed in culture, depending on the age of the donor. Human ES cells, because they are completely uncommitted to any cell type, display a property of our reproductive cells such that they replicate without limit (a property called cellular immortality). If sufficient conditions for growth were provided, within months the cells would overflow the limits of our solar system. This property of the cells is, as far as we know, unique to these cells. It is the reason babies are born young, and any cell derived from human ES cells would generally be expected to therefore be "young." In addition, based on data obtained from nonhuman animal ES cells, we should expect that human ES cells are more amenable to precise genetic modification (a process known as homologous recombination).

In addition, human ES cells display the unique property of totipotency, or near-totipotency. Human ES cells are unique in that they stand near the base of the developmental tree. These cells are therefore frequently designated "totipotent" stem cells, meaning that they are potentially capable of forming any cell or tissue type needed in medicine. In addition to forming any cell type, they are unique in their ability to self-assemble into complex multicellular tissues such as intestine, full thickness skin, kidney tissue, and so on. They differ from adult stem cells that are "pluripotent" that is, capable of forming several, but only a limited number, of cell types. An example of pluripotent adult stem cells are the bone marrow stem cells now widely used in the treatment of cancer and other life-threatening diseases.

These two unique characteristics of human ES cells open the door to manifold novel therapeutic strategies. It may not be an exaggeration to state that the combination of the ability to precisely genetically modify these cells by making any number of targeted genetic modifications and the ability to make any cell type may have as profound an application in medicine as the ability to arrange electrical components has made in the electronics industry.

It would be tragic if federal funding were generously applied to the new technologies of human ES cells and no parallel vision was applied to the problem of histo-compatibility. Human ES cells obtained from IVF preimplantation embryos are not identical to the patient, that is they are "allogeneic". We should expect that such cells would be rejected by the patient's immune system.

The primary purpose in funding human ES cell research is not just the pure pursuit of human knowledge, but rather to accelerate the delivery of novel therapeutics to afflicted people. We must address from the beginning how we are going to make these cells useful in transplantation. Several approaches can be envisioned to solve the problem of histocompatibility. One approach would be to make vast numbers of human ES cell lines that could be stored in a frozen state. This "library" of cells would then offer varied surface antigens, such that the patient's physician could search through the library for cells that are as close as possible to the patient. But these would likely still require simultaneous immunosuppression that is not always effective. In addition, immunosuppressive therapy carries with it increased cost, and the risk of complications including malignancy and even death.

Another theoretical solution would be to genetically modify the cultured ES cells to make them "universal donor" cells. That is, the cells would have genes added or genes removed that would "mask" the foreign nature of the cells, allowing the patient's immune system to see the cells as self. While such technologies may be developed in the future, it is also possible that these technologies may carry with them unacceptably high risks of rejection or other complications that would limit their practical utility in clinical practice.

The Use of Nuclear Transfer in Medicine

The recent success in the cloning of animals from various body cells demonstrates that the transfer of a body cell into the environment of an egg cell can "reprogram" the body cell back to an embryonic state. We have recently demonstrated that such technology actually rebuilds the replicative lifespan as well, suggesting that "young" cells can be derived from "old" cells. This is a profound development and perhaps the ideal solution for making real the longstanding dream of transplantation medicine; namely, to be able to offer any patient, even an aged patient, young healthy cells of any kind that are their own cells, not expected to be rejected by their immune system.

Much of the current public debate on stem cell therapy is focused on the appropriateness of federal funding for research on stem cells derived from IVF preimplantation embryos. While IVF-derived cells are useful for research and a limited number of clinical applications, it can play no significant role in the treatment of many common illnesses such as heart disease, renal failure, or diabetes. To regenerate diseased tissue with the complications associated with immunosuppression, cells with genomes identical to that of each patient will be required. IVF-derived stem cells, created by merging two unrelated genomes, cannot meet this requirement, but stem cells produced by culturing and reprogramming the patient's own mature tissue cells can. While this process, called therapeutic cloning, introduces another hot button term into the discussion, its product – specialized human tissue – is in all ways different from its morally suspect cousin, reproductive cloning. Shifting the debate away from IVF, which raises the unrelated issue of what to do with unwanted IVF tissue, and clarifying the distinction between therapeutic and reproductive cloning will better serve the public, science, and most of all, patients by focusing attention on the process as it will ultimately be employed.

The recent rush to legislate and ban all uses of cloning technology in the human species, including all applications in medicine are tragically shortsighted. With adequate debate, surely we are intelligent enough, and compassionate enough, to find an avenue that would allow researchers to ethically produce genetically matched ES cells for the millions afflicted with debilitating disease. A broad ban could set back critical research many years. In addition, it would shift the lead in this important area of research to other countries that have formally approved it such as the United Kingdom. The current emotional reaction to the uses of cloning technology is not unlike the early emotional reaction to in vitro fertilization that was labeled by many as "test tube babies". There was a similar rush to propose legislation banning test tube babies in the 1970s with many ethical leaders, including many of the same leaders that oppose therapeutic uses of nuclear transfer, calling test tube babies "illicit" and "immoral". Today, I believe we have a consensus that "test tube babies" are actually pro-life, and pro-family and a compassionate and wise use or modern technology. How tragic for many families it would have been if we had rushed to ban this important area of research in the heat of the moment.

Legal Considerations

Biotechnology is widely acknowledged to be a valuable growth industry for the United States. It has spawned some of the greatest advances in recent medical research. For example, biotechnology was the first to sequence the human genome leading to what is likely to be hundreds of novel medical therapies. In addition to these benefits accrued to those afflicted with disease, the biotechnology industry is not an insignificant source of employment, having created nearly a million jobs in the United States.

Patents are the lifeblood of the biotechnology industry. While many high tech companies benefit from patents in the battle against unfair competition, the issue is even more critical for the biotechnology industry. Unlike the computer industry, for example, the timeline for product development in biotechnology is often 10-15 years from conception to product on the shelf. The risks associated with novel therapeutic development combined with the long timelines usually makes patents an essential component for investors. Without patents to insure that success in product development will allow a window of protection from those that merely copy the technology at no cost, investors simply would not expend the monies necessary for product development. Indeed, there are many compounds that would have significant human benefit that are not developed and approved by the FDA simply because they are not patentable.

The human ES cell is currently patented under two issued U.S. patents. The first, U.S. Patent No. 5,843,780 issued on December 1, 1998 is based on earlier work in nonhuman primates, however, the claims encompass the human species. The second patent, U.S. Patent No. 6,200,806 issued on March 13, 2001 specifically cover the human species. It has been suggested that as a matter of principle, cultures of human ES cells should not be patentable subject matter. One objection often voiced is that life forms, such as living cells, especially those that have not been in any way modified, should not be patentable. The subject of the patentability of living organisms has been addressed by the Supreme Court. In the landmark June 16, 1980 decision in the case of Diamond v. Chakrabarty, the Chief Justice stated that living organisms were patentable under U.S. law:

[T]he patentee has produced a new bacterium with markedly different characteristics from any found in nature and one having the potential for significant utility. His discovery is not nature's handiwork, but his own; accordingly, it is patentable subject matter..." (Diamond v. Chakrabarty, 447 U.S. 303,310,206 USPQ 193,197 (1980)

But the history of patents on isolated living organisms goes back much further. One of the first U.S. patents issued on a living organism was granted in 1873 to Louis Pasteur that included a claim to a biologically pure culture of yeast as a composition of matter. Claim two of U.S. patent 141,072 read: Yeast, free from organic germs of disease, as an article of manufacture."

Another objection that arises uniquely in the context of human ES cells is that the patenting of human ES cells is the patenting of human life. The current position of the U.S. Patent and Trademark Office appears to be that humans are not patentable subject matter. As stated by Donald J. Quigg, Assistant Secretary and Commissioner of the USPTO on April 7, 1987:

"A claim directed to or including within its scope a human being will not be considered to be patentable subject matter under 35 U.S.C. 101. The grant of a limited, but exclusive property right in a human being is prohibited by the Constitution. Accordingly, it is suggested that any claim directed to a non-plant multicellular organism which would include a human being within its scope include the limitation 'non-human' to avoid this ground of rejection . . . . To the extent that the claimed subject matter is directed to a non-human 'nonnaturally occurring manufacture or composition of matter--a product of human ingenuity' (Diamond v. Chakrabarty [447 U.S. 303 (1980)], such claims will not be rejected under 35 U.S.C. 101. as being directed to nonstatutory subject matter."

Advanced Cell Technology's Response to Resent Announcements on Human Cloning

The announcement by Dr. Brigitte Boisselier, CEO of Clonaid, and some others, concerning the birth of a human clone, whether true or not has brought the discussion of reproductive and therapeutic cloning back into the forefront of the news. Advanced Cell Technology believes:

1. That the use of somatic cell nuclear transfer (SCNT), "cloning" technology for the purpose of human reproduction is wrong and condemns it based on ethical, moral and scientific reasons and is in agreement with the overwhelming majority of the scientific, medical, moral community and general population in condemning attempts to produce a human clone.

2. That congress should pass legislation that would prohibit using SCNT or "cloning" technology for reproducing humans, while permitting medical or therapeutic research which could save millions of human lives and improve the quality of life for millions more. We support Senate bill (S. 24392) which has bipartisan support and is cosponsored by Senators Arlen Spector, Dianne Feinstein, Orrin Hatch, and Edward Kennedy.

3. That the U.S. should play a leadership role at the UN in trying to outlaw reproductive cloning and permitting therapeutic cloning. It is urgent that the UN act while only four groups (Dr. Severino Antinori, Dr. Panayiotis Zavos, Dr. Richard Seed, and Dr. Brigitte Boisselier), have publicly expressed

an interest in producing a human clone. The technology exists and with enough time and money, a human clone will be produced unless actions are taken to prevent it across the globe. There is a strong need to eliminate the countries of refuge for the fringe scientists who desire to create a human clone.

4. That it is acceptable to extract stem cells from days old cloned cells, cultured in the laboratory for regenerative medical purposes. SCNT technology has advanced to the point that stem cells can be extracted from a microscopic ball of cells numbering between 80 and 100. This has all but eliminated any opposition from bioethicists and scientists concerning the use of pre-implantation embryos. As such, this promising research is supported by the larger scientific, medical, moral and general population, unlike its counterpart, human reproductive cloning. In terms of moral, scientific and medical support, the two have nothing in common.

5. That scientific research continues to demonstrate that embryonic stem cells show significantly more promise than adult stem cells. As such, embryonic stem cell research must continue and we should not let the actions of a few rogue scientists distract, delay or lead to the prohibition of that research.

Those interested in seeing therapeutic cloning continue but the prevention of human cloning, should ask their congressional representatives to support Senate bill 24392.

**Testimony of Michael D. West, Ph.D. President & CEO
Advanced Cell Technology, Inc. before the
Subcommittee on Labor, Health and Human Services,
Education and Related Agencies of the Senate
Committee on Appropriations, December 4, 2001**

Mr. Chairman and members of the Subcommittee, my name
is Michael D. West and I am the President and Chief
Executive Officer of Advanced Cell Technology, Inc., a
biotechnology company based in Worcester,
Massachusetts. A copy of my curriculum vitae is presented
in Appendix A.

INTRODUCTION

I am pleased to testify today regarding human embryonic
stem cell and nuclear transfer technology and their
applications in medicine. I would like to first speak to the
potential benefits of this emerging science, and then speak
to the objections that opponents have raised.

The Potential Benefits of ES and NT Technology

Human ES cells are unique in the history of medical
research for at least two reasons. First, they alone are
totipotent stem cells. By stem cells, we mean cells that can
branch out like the stems of a tree, becoming other cell
types. By "totipotent" we mean to say that they stand near
the base or "trunk" of the developmental tree and so are
capable of forming any cell or tissue type needed in

medicine. In addition to forming any cell type, they are unique in their ability to self-assemble into complex multicellular tissues such as intestine, full thickness skin, kidney tissue, and so on. They differ in this respect from adult stem cells that are "pluripotent" that is, capable of forming several, but only a limited number, of cell types. One can think of adult stem cells as limbs further out on the branches of a tree. While able to branch out in several different directions, only the trunk of the tree branches out into very leaf and limb. An example of adult pluripotent adult stem cells are the bone marrow stem cells now widely used in the treatment of cancer and other life-threatening diseases.

The second distinguishing feature of ES cells is the ease with which they can be purposefully modified in a precise manner. This precise genetic modification is designated "gene targeting". The enhanced ability of ES cells to be modified with precision opens the door to likely many hundreds of clinical applications making human cells of any kind, genetically modified in any way.

These two unique characteristics of human ES cells open the door to manifold novel therapeutic strategies. It may not be an exaggeration to state that the combination of the ability to precisely genetically modify these cells by making any number of targeted genetic modifications and the ability to make any cell type may have as profound an application in medicine as the ability to arrange electrical components has made in the electronics industry.

To attempt to name every disease potentially impacted through this technology would require a larger report. A few examples would be to manufacture neurons for degenerative diseases such as Parkinson's and spinal cord injury. Gene targeting to find and correct mutations could be used to manufacture neuronal stem cells for childhood retardation from diseases like Rett syndrome. Heart and skeletal muscle cells could be used for heart failure and age-related skeletal muscle wasting, and targeted genetic modification could be useful in muscular dystrophy. Blood forming cells would be useful in bone marrow grafting after cancer treatments, and anemias. Precision genetic modification could lead to better therapies for inherited blood cell disorders such as sickle cell anemia and infectious diseases such as AIDS.

I would argue that the debate over the number of human ES stem cell lines approved for federal funding largely misses the point. Human ES cells obtained from IVF preimplantation embryos are not identical to the patient, that is they are "allogeneic". We should expect that such cells derived from the 20-60 approved lines would be rejected by the patient's immune system. The primary purpose in funding human ES cell research is not just the pure pursuit of human knowledge, but rather to accelerate the delivery of novel therapeutics to afflicted people. We must address from the beginning how we are going to make these cells useful in transplantation. Several approaches can be envisioned to solve the problem of histocompatibility. One approach would be to make vast numbers of human ES cell lines that could be stored in a frozen state. This "library" of cells would then offer varied surface antigens, such that the patient's physician could search through the library for cells that are as close as possible to the patient. But these would likely still require simultaneous immunosuppression that is not always effective. In addition, immunosuppressive therapy carries with it increased cost, and the risk of complications including malignancy and even death.

Another theoretical solution would be to genetically modify the cultured ES cells to make them "universal donor" cells. That is, the cells would have genes added or genes removed that would "mask" the foreign nature of the cells, allowing the patient's immune system to see the cells as self. While such technologies may be developed in the future, it is also possible that these technologies may carry with them unacceptably high risks of rejection or other complications that would limit their practical utility in clinical practice.

The Use of Nuclear Transfer in Medicine

The recent success in the cloning of animals from various body cells demonstrates that the transfer of a body cell into the environment of an egg cell can "reprogram" the body cell back to an embryonic state. We have recently demonstrated that such technology actually rebuilds the replicative lifespan as well, suggesting that "young" cells can be derived from "old" cells. This is a profound development and perhaps the ideal solution for making real the longstanding dream of transplantation medicine; namely, to be able to offer any patient, even an aged patient, young healthy cells of any kind that are their own cells, not expected to be rejected by their immune system.

Nuclear transfer offers an important solution of the problem of tissue rejection. The procedure would involve the patient donating living cells to a physician, who would then reprogram them back to a totipotent state using the cloning procedure. This is called *therapeutic cloning*, to distinguish it from *reproductive cloning* which is designed to clone an entire human being. The cells and tissues made from these cloned stem cells would be expected to be grafted stably for the life of the patient without immunosuppression.

Answers to the Opponent's Objections:

1). The preimplantation embryo is a human life and to use therapeutic cloning is to "clone and kill".

Answer: A preimplantation embryo is human cellular life, but not a human life. The trillions of cells in our body are all truly alive. Therefore human cells growing in a laboratory dish would rightly be called human cellular life, but no reasonable person would say that they are "a human life". In the first few days following the fertilization of an egg cell by a sperm cell, there develops a microscopic ball of cells called a preimplantation embryo. This embryo is destined to die unless it implants in the uterus to form a pregnancy. Indeed, it is estimated that 50-80% of these preimplantation embryos naturally formed in a woman's body never implant and therefore die. A human life, as opposed to simply cellular life begins, at the earliest, at or around day 14 of human development at around the time the preimplantation embryo attached to the uterine wall in the mother. Prior to day 14, the preimplantation embryo has no body cells of any kind, and, in fact, has no cells even committed to somatic cell lineages. Indeed, the embryo has not individualized. Once this ball of cells attaches to a uterus, one or even two or more individuals can form from it. In addition, if it merges with another preimplantation embryo and then forms a pregnancy, it will become just some of the cells in the resulting person. Therefore, many have concluded that at this early stage before a pregnancy, the preimplantation embryo has not individualized and therefore it would be illogical to attribute to it even the earliest status of personhood.

2). Therapeutic cloning is merely theoretical; there is no reason to suggest it will work.

Answer: Those that make this objection appear to be simply uninformed of the scientific literature. There are published reports of success of therapeutic cloning research in at least two mammalian species; namely mice (1-2). While never performed in a human, the animal data suggests that therapeutic cloning has great promise. The National Academy of Sciences has formally recommended in a report titled "Stem Cells and the Future of Regenerative Medicine" as follows:

"Recommendation: In conjunction with research on stem cell biology and the development of potential stem cell therapies, research on approaches that prevent immune rejection of stem cells and stem cell-derived tissues should be actively pursued. These scientific efforts include the use of a number of techniques to manipulate the genetic makeup of stem cells, including somatic cell nuclear transfer.[3]"

3). Allowing therapeutic cloning would cause a "slippery slope" effect, whereby regulating human reproductive cloning would not be possible.

Answer: In reality the procedures to clone a human being are well known in the scientific literature. The widespread use of therapeutic cloning would not significantly increase the likelihood of the success of an effort to clone a human being. In addition, laws can easily be written to allow one and prohibit the other as reproductive cloning requires the transfer of a cloned preimplantation embryo into a uterus.

4). Therapeutic Cloning will lead to "embryo farms", "Nazi-like experimentation", a "Brave New World", etc.

Answer: The opponents of many recent medical technologies have resorted to such inflammatory language in the absence of a rational basis of objection. Therapeutic cloning guidelines could easily be constructed to limit development to less than 14 days as is the current practice of in vitro fertilization.

Summary

In conclusion, nuclear transfer and human embryonic stem cell technology offer novel pathways to develop lifesaving therapies that will impact the lives of millions suffering from such diseases as Parkinson's disease, diabetes, arthritis, heart disease, kidney failure, spinal cord injury, liver failure, skin burns, blood cell cancers, to name only a few. The gravity of this issue calls for a compassionate, reasoned, and dispassionate debate. History will judge us harshly if we as a society fail to recognize and deliberate carefully upon a medical technology that could so powerfully alleviate the suffering of our fellow human being.

References

1.   Cibelli, J. B., Stice, S. L., Golueke, P. J., Kane, J. J., Jerry, J., Blackwell, C., Ponce de Leon, F. A. & Robl, J. M. (1998) *Nat Biotechnol* **16,** 642-6.

2.   Wakayama, T., Tabar, V., Rodriguez, I., Perry, A. C., Studer, L. & Mombaerts, P. (2001) *Science* **292,** 740-3.

3.   *Stem Cells and the Future of Regenerative Medicine* - National Academy Press (2001), Washington, p39.

Bibliography of Congressional Testimony of Dr. Michael West

See http://www.gpo.gov/fdsys/search/home.action

S. Hrg. 105-939 - STEM CELL RESEARCH [PDF 824 KB]
Congressional Hearings. General. Appropriations.
Wednesday, December 2, 1998.
...16 Statement of **Michael West**, Ph.D., president and chief executive...Gearhart, Dr. James Thomson, and Dr. **Michael West** to join us at the panel. Our first...modifying the human germ line. STATEMENT OF **MICHAEL WEST**, Ph.D., PRESIDENT AND CHIEF...More Information

S. Hrg. 107-499 - STEM CELLS, 2001 [PDF 1022 KB]
Congressional Hearings. General. Appropriations.
Wednesday, July 18, 2001.
...Now we turn to Dr. **Michael West**, president and CEO...committee. My name is **Michael West**. I am the president...I was the founder of **Geron** Corporation in Menlo...will turn now to Dr. **Michael West**, president and...More Information

S. Hrg. 105-939 - STEM CELL RESEARCH [PDF 824 KB]
Congressional Hearings. General. Appropriations.
Wednesday, December 2, 1998.
...16 Statement of **Michael West**, Ph.D., president and...research and development, **Geron** Corp...human germ line. STATEMENT OF **MICHAEL WEST**, Ph.D., PRESIDENT AND...Specter. We now turn to Dr. **Michael West**, president and chief...More Information

S. Hrg. 108-844 - CLONING: A RISK TO WOMEN?
[PDF 1901 KB]
Congressional Hearings. General. Commerce,
Subcommittee on Science, Technology and Space.
Thursday, March 27, 2003.
...Dr. Thomas Okarma, Chief Executive of **Geron**
Corporation, to state that it would take...Report No. 19;
Series 23, 1997. \2\ **Geron** Inc. Advanced Cell Technology.
\3...issue, I would note that 18 months ago, **Michael West**,
of Advanced Cell Technology, at a...More Information

147 Cong. Rec. H4916 - HUMAN CLONING
PROHIBITION ACT OF 2001 [PDF 252 KB]
Congressional Record. Regarding H.R. 2505. Mr.
SENSENBRENNER, Mr. CONYERS, and others.
Tuesday, July 31, 2001.
...Green of Dartmouth University, and **Michael West**,
Robert Lanza, and Jose Cibelli of...emphasis added] Earlier
this month, **Michael West**, the head of the major biotech
firm...Green of Dartmouth University, and **Michael West**,
Robert Lanza, and Jose Cibelli...More Information

147 Cong. Rec. (Bound) 15203 - Mr. CONYERS. Mr.
Speaker, I yield such time as he may consume to the
gentleman from New York (Mr... [PDF 231 KB]
Congressional Record (Bound Edition). Regarding H.R.
214. Mr. CONYERS, Mr. NADLER, and others. Tuesday,
July 31, 2001.
...Green of Dartmouth University, and **Michael West**,
Robert Lanza, and Jose Cibelli of...emphasis added] Earlier
this month, **Michael West**, the head of the major biotech
firm...Green of Dartmouth University, and **Michael West**,
Robert Lanza, and Jose Cibelli...More Information

149 Cong. Rec. H1397 - HUMAN CLONING
PROHIBITION ACT OF 2003 [PDF 294 KB]

Congressional Record. Regarding H.R. 105. Mrs. MYRICK, Mr. McGOVERN, and others. Thursday, February 27, 2003.

...Green of Dartmouth University, and **Michael West**, Robert Lanza, and Jose Cibelli...cloning, even though the chairman of **Geron**, Thomas Okarma, is quoted on...Green of Dartmouth University, and **Michael West**, Robert Lanza, and Jose...More Information

148 Cong. Rec. S7862 - IN MEMORY OF TIMOTHY WHITE [PDF 114 KB]
Congressional Record. Mr. HATCH and Mr. SMITH of New Hampshire. Thursday, August 1, 2002.
...among both British and American researchers. Last December, **Michael West** of Advanced Cell Technology predicted that within 6 months...by any measure. Thomas Okarma, the chief executive of **Geron** Corp., a cell therapy company, has no interest in...More Information

S. Hrg. 107-499 - STEM CELLS, 2001 [PDF 1022 KB]
Congressional Hearings. General. Appropriations. Wednesday, July 18, 2001.
...Statement of **Michael West**, Ph.D...and CEO, **Advanced Cell Technology**...School; **Michael West**, Ph.D...and CEO of **Advanced Cell Technology**. Let me...turn to Dr. **Michael West**, president and CEO of **Advanced Cell Technology**. Dr....More Information

S. Hrg. 105-939 - STEM CELL RESEARCH [PDF 824 KB]
Congressional Hearings. General. Appropriations. Wednesday, December 2, 1998.
...Statement of **Michael West**, Ph.D...officer, **Advanced Cell Technology**...STATEMENT OF **MICHAEL WEST**, Ph.D...OFFICER, **ADVANCED CELL TECHNOLOGY**

Senator Specter...was Dr. **Michael West**, of **Advanced Cell Technology**, who...<u>More Information</u>

<u>S. Hrg. 107-444 - CLONING, 2001 [PDF 274 KB]</u>
Congressional Hearings. General. Appropriations. Tuesday, December 4, 2001.
...years ago, Dr. **Michael West** of **Advanced Cell Technology** testified before...be Dr. West, **Michael West**, Dr. Ronald...appreciate that. Dr. **Michael West**, we'll start...president and CEO of **Advanced Cell Technology** in...<u>More Information</u>

<u>147 Cong. Rec. H4916 - HUMAN CLONING PROHIBITION ACT OF 2001 [PDF 252 KB]</u>
Congressional Record. Regarding H.R. 2505. Mr. SENSENBRENNER, Mr. CONYERS, and others. Tuesday, July 31, 2001.
...University, and **Michael West**, Robert Lanza...Jose Cibelli of **Advanced Cell Technology**-- confirmed...this month, **Michael West**, the head of...biotech firm **Advanced Cell Technology** (ACT) of Worcester...University, and **Michael West**, Robert Lanza...Jose Cibelli of **Advanced Cell Technology**--
......<u>More Information</u>

<u>147 Cong. Rec. (Bound) 15203 - Mr. CONYERS. Mr. Speaker, I yield such time as he may consume to the gentleman from New York (Mr... [PDF 231 KB]</u>
Congressional Record (Bound Edition). Regarding H.R. 214. Mr. CONYERS, Mr. NADLER, and others. Tuesday, July 31, 2001.

...University, and **Michael West**, Robert Lanza...Jose Cibelli of **Advanced Cell Technology**-- confirmed...this month, **Michael West**, the head of...biotech firm **Advanced Cell Technology** (ACT) of

Worcester...University, and **Michael West**, Robert Lanza...Jose Cibelli of **Advanced Cell Technology**-- ......More Information

S. Hrg. 107-812 - HUMAN CLONING: MUST WE SACRIFICE MEDICAL RESEARCH IN THE NAME OF A TOTAL BAN? [PDF 1142 KB]
Congressional Hearings. General. Judiciary. Tuesday, February 5, 2002.
...and human cloning, **Michael West** of **Advanced Cell Technology** (ACT) predicted...technique used by **Advanced Cell Technology**? Mr. Greely. Can...produce stem cells, **Advanced Cell Technology** in...More Information

S. Hrg. 108-844 - CLONING: A RISK TO WOMEN? [PDF 1901 KB]
Congressional Hearings. General. Commerce, Subcommittee on Science, Technology and Space. Thursday, March 27, 2003.
...is now known that **Advanced Cell Technology** of Massachusetts...2\ Geron Inc. **Advanced Cell Technology**. \3\ David Stevens...that 18 months ago, **Michael West**, of **Advanced Cell Technology**, at a Senate...More Information

149 Cong. Rec. H1397 - HUMAN CLONING PROHIBITION ACT OF 2003 [PDF 294 KB]
Congressional Record. Regarding H.R. 105. Mrs. MYRICK, Mr. McGOVERN, and others. Thursday, February 27, 2003.
...researchers at **Advanced Cell Technology** in Massachusetts...University, and **Michael West**, Robert...Cibelli of **Advanced Cell Technology**-- wrote...University, and **Michael West**, Robert...Cibelli of **Advanced Cell**...More Information

148 Cong. Rec. S7862 - IN MEMORY OF TIMOTHY WHITE [PDF 114 KB]
Congressional Record. Mr. HATCH and Mr. SMITH of New Hampshire. Thursday, August 1, 2002.
...said to be "falling from favor" among both British and American researchers. Last December, **Michael West** of **Advanced Cell Technology** predicted that within 6 months, his company would be ready to create "magic" cells that...More Information

S. Hrg. 107-541 - CLONING, 2002 [PDF 404 KB]
Congressional Hearings. General. Appropriations. Thursday, January 24, 2002.
...practically every week. Three years ago, Dr. **Michael West** of **Advanced Cell Technology** testified before this committee about a new...mention a few. Well, late last year Dr. **Michael West** announced that he had taken the first...More Information

No Genetically Modified Food or Humans?

I can see areas that gravely concern me. The ability to genetically modify these cells could eventually lead to genetically modified human beings, in any way. The concept of beginning to genetically engineer the human being, is in a sense playing god with human life. These types of technologies could be adopted to such ends. The intent is not to cure disease but to enhance human characteristics. There are thoughtful people here in the United States who say we are going to be doing this. If your children are going to be competing with other children who have been genetically engineered to have enhanced memory, you are going to want your children to have that too. Those are ethical debates that are worthy of discussion. I don't think it's imminent, but those issues trouble me as an individual.

Dr. Michael West, President and CEO BioTime "The Stem Cell Controversy"

People may be familiar with the debate around the genetic modification of food and its labeling, but I wonder if people know there is not the same level of discussion about the genetic modification of human beings.[25] In 2003, prior to Dr. West publishing his book, "The Immortal Cell", I urged him to provide ethical guidelines for the future of research in this area. I believe it is past due, as scientists need to know what we as a society view as the limits. I don't believe it is fair to address the issue after the fact, when we read a headline of a scientist doing something we would prefer they don't.

---

[25] There is debate if gene editing that results in changes in food is the same as genetic modification.

During the debate on the use of embryonic stem cells and cloning, it was argued using embryonic stem cells and cloning techniques for therapeutic cloning was ok, but not for reproductive cloning. It was argued that it should be illegal to use those techniques to create a human being. What was not discussed was using those techniques to *modify* a human organism. Earlier this year, headlines announced that scientists had produced the first genetically modified humans and had been doing it for a while.[26] [27] At the Institute for Reproductive Medicine and Science of St. Barnabas in New Jersey, they announced that at least 30 healthy babies were born as a result of an experiment designed to help infertile women conceive. The experiment consisted of inserting extra genes from a female donor into the eggs of women who struggled with infertility, before their eggs were fertilized. When the DNA of these babies was tested, it showed they had three parents, not two. What is disconcerting is this procedure resulted in a change in the germ line such that all future offspring will potentially have this characteristic.

But this is not the only such case. Researchers at Oregon Health and Science University want permission to use "mitochondrial manipulation technologies." The procedures involve removing the nuclear material either from the egg or embryo of a woman with inheritable mitochondrial disease and inserting it into a healthy egg or embryo of a donor whose own nuclear material has been discarded.

Mostly the result of a genetic defect, approximately 1,000 – 4,000 children in the US develop mitochondrial disease,

---

[26] World's first GM babies born by Michael Hanlon, Daily Mail, http://www.dailymail.co.uk/news/article-43767/Worlds-GM-babies-born.html.
[27] Genetically Modified Babies, by Marcy Darnovsky Feb. 23, 2014.

which typically prevent mitochondria from converting food into energy. The symptoms can range from mild to devastating. What the developer of the technique said was this was a "mere" "tweaking of the reproductive process." Rightly understood and defined, this is not technically genetic modification, or genetic engineering, as there was no "splicing and dicing" of genes." They did not use recombinant methods to create something that does not exist in nature. As a result, some might feel quite comfortable with the use of this technology, while others don't. Given the powerful tools that exist today, what they did was mere child's play compared to what they could do. What we want to know is, ***"what are the guidelines, what should they be allowed to do and forbidden to do?"***

In order to understand the problem we face, it requires some basic understanding. What we now call genetic modification, was at one time called genetic programming

Let's start by defining Regenerative Medicine.

Regenerative medicine is a branch of <u>translational research</u>[1] in <u>Tissue Engineering</u> and <u>Molecular Biology</u> which deals with the "process of replacing, engineering or regenerating human cells, tissues or organs to restore or establish normal function".[2] This field holds the promise of engineering damaged tissues and organs via stimulating the body's own repair mechanisms to functionally heal previously irreparable tissues or organs.[3]

Regenerative medicine also includes the possibility of growing tissues and organs in the laboratory and safely implanting them when the body cannot heal itself. This can potentially solve the problem of the shortage of organs available for donation, and the problem of <u>organ transplant</u>

rejection if the organ's cells are derived from the patient's own tissue or cells.[4][5][6]

Widely attributed to having first been coined by William Haseltine (founder of Human Genome Sciences),[7] the term "Regenerative Medicine" was first found in a 1992 article on hospital administration by Leland Kaiser. Kaiser's paper closes with a series of short paragraphs on future technologies that will impact hospitals. One such paragraph had "Regenerative Medicine" as a bold print title and went on to state, "A new branch of medicine will develop that attempts to change the course of chronic disease and in many instances will regenerate tired and failing organ systems."[8][9]

Regenerative medicine refers to a group of biomedical approaches to clinical therapies that may involve the use of stem cells.[10] Examples include the injection of stem cells or progenitor cells (cell therapies); the induction of regeneration by biologically active molecules administered alone or as a secretion by infused cells (immunomodulation therapy); and transplantation of *in vitro* grown organs and tissues (tissue engineering).[11][12]

Information and footnotes from Wikipedia.

Another way to define regenerative medicine is as follows:[28]
In 2006, the US National Institutes of Health defined regenerative medicine as "the process of creating living, functional tissues to repair or replace tissue or organ function lost due to age, disease, damage, or congenital

---

[28] A good resource is *Regenerative Medicine.* Department of Health and Human Services. August 2006. </info/scireport/2006report.htm> http://stemcells.nih.gov/info/scireport/pages/2006report.aspx reference.

defects," if any of these processes involve the use of stem cells.

Here is another definition of regenerative medicine:

Regenerative medicine is a broad definition for innovative medical therapies that will enable the body to repair, replace, restore and regenerate damaged or diseased cells, tissues and organs. Scientists worldwide are engaged in research activities that may enable repair of damaged heart muscle after a heart attack, replacement of skin for burn victims, restoration of movement after spinal cord injury and regeneration of pancreatic tissue to produce insulin for people with diabetes. Regenerative medicine promises to extend healthy life spans and improve the quality of life by supporting and activating the body's natural healing.[29]

As we age, cells lose their ability to replicate and as a result this leads to various types of tissue or organ damage and loss of function and symptoms. There are few effective ways to treat the root causes of many age related diseases, injuries and congenital conditions. In many cases, clinicians can only manage patients' symptoms using medications or devices. It is believed as much as 75% of health care expenses are for degenerative chronic disease. This is a real problem as the baby boomer population ages. As a result we can expect a significant growth in age related diseases. Regenerative medicine may hold the key to solving this pending health care crisis because it is a way to make new tissues or organs or repair or replace damaged ones. Regenerative medicine is a way to regenerate cells.

---

[29] From Pall corporation web site
http://www.pall.com/main/medical/frequently-asked-questions-cell-therapy-38848.page.

Regenerative medicine depends upon embryonic stem cells (even though the stem cells themselves do not come from embryos) to make the cells necessary for tissue or organ repair or replacement. These stem cells have the ability to become any type of cell or tissue the body may need. It is similar to having an old car where the engine is worn out after two million miles and replacing the old engine with a new one with zero miles on it. As long as it is the same make and model as the one it is replacing it will work just fine. This type of regenerative medicine requires stem cells, but that is no longer an issue as we have the ability to make stem cells without destroying embryos. Because we are using the patient's own cells, we can make an exact replica of the heart or engine we are replacing so the body does not reject it.

In addition to these stem cells having the ability to become any type of cell in the body, they can be modified through genetic engineering. We have the ability to genetically engineer or modify any stem cell. This provides great potential for treatment. While this technique could be used for good, we have not had a discussion as to its ethical limits.
Regenerative medicine also requires another technique, therapeutic cloning to develop the millions of cells we need for this type of therapy. When cloning is used, it does not result in a copy or duplicate of an old cell, but rather a new cell with no "miles" on it. Millions of these cells can be copied and as a result new therapies can be developed. Again the ethics of this technique have largely been settled – therapeutic cloning is ok, but reproductive cloning is problematic from an ethical perspective.

Some life extensionists, or anti-aging scientists suggest that therapeutic cloning and stem cell research could one day provide a way to generate cells, body parts, or even entire bodies (generally referred to as <u>reproductive cloning</u>) that would be genetically identical to a prospective patient. Recently, the US Department of Defense initiated a program to research the possibility of growing human body parts on mice.[64]

To understand why regenerative medicine is so revolutionary, we will look at two letters Dr. Michael West president of BioTime wrote to his shareholders:

Regenerative medicine: a revolutionary means of treating disease

The field of regenerative medicine began with the search for a means to rebuild tissues afflicted with chronic degenerative disease. Currently some 75% of health care costs are associated with chronic diseases, and those costs are expected to grow significantly in the coming years as a result of the approaching tsunami of aging baby boomers. Unlike mechanics who can repair a machine by replacing broken components, medical science never had an effective means of producing replacement tissues from the multitude of cell types in the human body, until now. Cells and tissues available from organ donation are often not suitable for use due to potential transplant rejection.

Given the large unmet need for cell and tissue replacement therapies to treat degenerative diseases, an effort was organized in 1995 to build a new class of medical therapies based on pluripotent stem cells capable of proliferating without limit. The discovery of these stem cells brought into focus the possibility of developing industrial-scale means of producing various cell types of the human body that could be used to replace tissues lost as a result of the onset of degenerative diseases. In the late 1990s, it also became clear that it was going to become possible to make such cells genetically identical to those of a patient on an affordable basis through what are called "reprogramming technologies". These new technologies provide the opportunities of manufacturing all of the cell types of the human body on an industrial scale and matching those cells, when necessary, to the patient to eliminate transplant rejection. This new approach has the potential to provide therapies for some of the most significant unsolved problems in medicine, such as heart failure, in which the loss of heart muscle cells leads to a weakened heart. Many other age-related chronic degenerative diseases could potentially be treated with these novel technologies, including Parkinson's disease, osteoarthritis, osteoporosis, age-related macular degeneration, and atherosclerosis, to name a few.

The stem cells that have caused this excitement are of two types: human embryonic stem (hES) cells and induced pluripotent stem (iPS) cells. hES cells are cultures of continuously proliferating cells (cell lines), and were originally derived from clusters of cells resulting from surplus fertilized egg cells produced through in vitro fertilization procedures and that were designated by the donors to be discarded. iPS cells are cells produced from a donated cell of the body, such as skin cells, and then coaxed back to an hES-like state by means of genetic modification. While hES cells are seen as an "off-the-shelf" approach to making all the cells of the body, iPS cells are seen as a means of making all of these cells genetically matched to a patient in cases in which transplant rejection would otherwise be a problem.

As revolutionary as hES and iPS cells are, it became clear during recent years that the promise of these cells is intimately linked to a significant hurdle. Put quite simply, these cells really do make all of the cells of the body, and they tend to display this immense power even when it is undesirable, such as when scientists are trying to generate only one cell type for therapeutic use. As a result, many different cell types that form a human being were arising in laboratory dishes, even when only one cell type was desired, and scientists had no effective means of sorting these cells in order to manufacture a medical-grade product of a single cell type. While scientists quickly published hundreds of scientific reports describing methods of making the many diverse cell types of the body—as you may have read—the biotechnology industry faced the tall hurdle of making cells in a purified and fully identified state, under Good Manufacturing Practices (GMP) standards, while simultaneously scaling cell production up to a volume sufficient to treat millions of patients. As the biotechnology industry and large pharmaceutical companies struggled with these issues, BioTime scientists were inventing entirely novel manufacturing technologies that we believe will redefine the industry.

The Starbucks Effect, Aging Boomers, and Solving the Health Care Crisis
JUNE 17, 2013 BY MICHAEL D. WEST, PH.D.

Conspicuously absent from the debate over the health care crisis is even a hint that modern science might actually come to the rescue. Perhaps everyone has already assumed that the problem is really just aging itself – and what can be done about that? While many of us baby boomers can remember the days of gathering in hamburger joints to brag about fixing up muscle cars, today we find ourselves sitting around in coffee shops lamenting that there's nothing to be done about age-related macular degeneration and the like. At the national level, we've also observed a giant shoulder shrug: in his 2013 State of the Union Address, President Obama claimed that "the biggest driver of our long-term debt is the rising cost of healthcare for an aging population." The fact that emerging medical innovations may offer some fixes for this crisis is all but missing from the current discourse.

Recently in the press was a scarcely noticed report about a team of researchers at the Oregon National Primate Research Center that, after over six years of work and over fifteen years of preparation in the field in general, has successfully generated stem cells by cloning. The promise of this technology unfortunately sounds alarms for some that mad scientists are about to regenerate a Hitler or a Stalin. In contrast, actual stem cell researchers see within the cloning process something almost magical in terms of its potential benefits for humankind. The novel methods we are witnessing emerge in our own time may provide pathways to a kind of cellular "time machine" that can generate young cells of any kind to replace diseased, aging cells. A sloughed skin cell, for example, could be provided at no harm by an older patient, and then the subsequent cloning process could be used to create young cells genetically identical to that patient's own cells, thereby circumventing the risk of transplant rejection associated with the current standard of care. This novel means of treating and perhaps curing numerous age-related degenerative diseases has come to be known as "therapeutic cloning," in order to distinguish it from "reproductive cloning," which refers to the more controversial use of cloning techniques to create a baby.

The recent report outlines how the authors managed to produce long sought-after rejuvenated stem cells by adding none other than caffeine to the cloning mix, among other tweaks made to the already established protocol. This "Starbucks effect" led to microscopic clusters of "pluripotent" stem cells capable of differentiating into an unlimited number of cell types. As a result, it is possible to create young cells of all tissue types without forming an actual cloned baby. This is the basis for referring to the tissue-specific approach as "therapeutic cloning," in order to distinguish the goal of the procedure from that of "reproductive cloning," which would be to create a complete human.

This line of research began about 15 years ago with the first isolation of human embryonic stem cells, which for the first time in the history of medicine opened the door onto a means of manufacturing, on an industrial scale, all of the cellular components of the human body. Scientists saw within this discovery a potential pathway to numerous therapies and cures, for example, the regeneration of new heart muscle to strengthen a failing heart and the generation of new, healthily functioning brain cells needed to treat Parkinson's disease. This emerging field of research and innovation has come to be known as regenerative medicine. With embryonic stem cell, nuclear transfer (cloning), and induced pluripotent stem cell technology all working, medical researchers now have multiple paths to designing these new therapies.

Among the many potential applications of regenerative medicine, likely the most important will be for treating age-related degenerative diseases. Cells throughout the body have tiny clocking mechanisms built into their DNA. As a result of this ticking-clock mechanism, cells gradually lose their ability to repair damage with the passage of time. One could argue that as our age expectancy has increased, so has the duration of our suffering: many people now experience years or even decades of chronic and debilitating disease. Consider also the financial stress when a family member is diagnosed with Alzheimer's, or has a stroke, or becomes blind due to macular degeneration, or has heart failure. We urgently need novel strategies such as those announced today in order to increase the quality of care for those who are suffering. Parallel to this need is the necessity of reducing the high costs of treating these increasingly prevalent diseases.

Unknown to the public at large, scientists finally have some impressive new tools to address both the financial and physical issues associated with age-related degenerative disease. If we were to mobilize our scientific community by funding a national discovery program to find cost-effective cures, we could combine and apply the efforts of our best minds and hands working within the emerging field of regenerative medicine. With such a synergistic program in place, we could potentially save our nation trillions of dollars over the coming decades and alleviate human suffering on an unprecedented scale. These goals are not only awe-inspiring, they are also potentially within reach. We should be encouraged to use these new discoveries in an intelligent and compassionate manner to cure degenerative diseases that have, throughout history, been considered as unavoidable, as our collective fate. Some day in the not-too-distant future, our thinking about aging itself may change radically and positively. Meanwhile, the burden of health care costs that our generation leaves to following generations will be mitigated substantially by the amelioration, even curing, of those diseases before they become so financially and physically costly. We owe our fellow man exploration in regenerative medicine. Moreover, such a program of discovery will be necessary if the United States desires to retain its leadership role in the world community.

FILED UNDER: <u>REGENERATIVE MEDICINE,</u>
<u>UPDATES FROM THE CEO</u>

The excitement with these new technologies is we can more effectively treat chronic diseases such as Parkinson's disease, osteoarthritis, osteoporosis, age-related macular degeneration, and atherosclerosis. We could treat genetic diseases such as sickle cell anemia and there are over 3,000 such genetic diseases. We could treat infectious diseases such as HIV by developing an HIV resistant immune system. Regenerative medicine provides vast potential to help treat major diseases, which would be a benefit to the world.

Where the potential ethical issue resides lies with the genetic engineering, or genetic modification or genetic programming of human embryonic stem cells. We now have or soon will have in some cases the ability to create enhancements or create entirely new human organisms. It is very similar to what we have experienced in the digital world by analogy.

In the digital world, all programming is based on a two-digit code of zeroes and ones. From that we have built new industries and destroyed old ones. Publishing has gone from analog to digital, but so have broadcasting, the movies, cable and the photography industry. We have recently experienced the result in digital technology when the barriers between industries were broken by digital technology. Previously, the

Telephone
Computer
Broadcast-- radio and TV
Movie
Video games
Publishing
Photography
Music

.....were all separate industries. At their base, it is a simple task of programming the [0,1's] to determine if you get a photograph, music or a software program. These industries converged as can be seen with smart phones, which contain some or all of these characteristics and more. Some smartphones have over a billion applications. The list goes on; there used to be companies that made printers, copiers, faxes and scanners, and they all converged. This shows walls between industries falling as a result of the digital revolution.

We can now (re)program stem cells, genes and DNA, the building blocks of life and as such we have the ability to code biologically the same way we code digitally. Just as we have various types of "coders" today, e.g. Java writing digital code we will have similar coders writing biological code for regenerative medicine and human genetic modification.

The current revolution in biology is based on four things. First, there is recombinant DNA technology, which provides scientists the opportunity to slice and dice DNA. This power makes it possible to arrange genes in new ways to manufacture new "products." The second is genomics, which provided scientists an opportunity to rapidly sequence and manipulate gene sequence information. Third, somatic-cell nuclear transfer (SCNT) – often referred to as cloning, either therapeutic or reproductive. This is what makes genetic modification possible. Fourth, the discovery of embryonic human stem cells, which is also critical to regenerative medicine and genetic modification.

To explain how all of this converges and makes for both a potentially exciting (or scary) future, we need to explain the basics. What these technologies do is provide us the tools.

Genetic programming, genetic modification is the convergence of these four technologies which allow us to write biological software, in a manner similar to writing digital software. Computer software is based on a binary system of zeros and ones. Although we use a higher programming language than writing zeros and ones, that is what is occurring at its base. With DNA, nature is using a quaternary system of G, C, A, T to program all of life. Genomics provides the map of what the genes do and where they go. It is one thing to know the code, it is another to have the ability to change it.

In biotechnology, there is an area of study called gene targeting. This area covers how to modify the code of life forms. Gene targeting is the addition or deletion of genes, for a limited or permanent period of time. This is done a number of ways:

"Knock-out" – this refers to knocking out a specific gene in a plant, animal or human species. In other words, when looking at the code or program, and you see a gene that is responsible for a particular characteristic, you can "knock-out" that gene.

A practical example is that pig organs are sometimes transplanted into humans. Since there is a shortage of organs, this is one way to solve the problem. However, the human body is trained to reject various types of foreign objects. As a result, for the pig organ to take, people have to have immunosuppression drugs. What if you could "knock-out" the gene in the pig that causes human rejection? That is in fact what they have done and are trying to improve. Other examples include knocking-out the gene that causes people to be allergic to cats, thus making an allergy free cat.

A plant example would be knocking-out the gene that produces caffeine.

Knocking-out a gene is like hitting the "delete" key in word processing. Using your genomic map, you find the section of code that you don't want and you delete it.

"Knock-in" – is when a gene is "knocked-in" to the sequence and as a result changes the code. Knocking in a gene is like hitting the "insert" key or "Find/Replace" on the key board. You have a piece of code you want to insert, because by inserting it you change the program in a way you desire.

Alteration – we have the ability to change one genetic letter to another for example, change an "A" to a "G" or a "C" to a "T"

Gene Splicing – is the cutting of one gene from one organism and pasting it into the DNA of another so that a characteristic can be transferred from one plant or animal to another.

Transgenic – of, relating to, or denoting an organism that contains genetic material into which DNA from an unrelated organism has been artificially introduced. The example of BioSteel in the introduction is such an example, where genes from a spider are inserted into a goat.

Cell Therapy – is when cellular material is injected into a patient. The FDA defines cell therapy as "The prevention, treatment, cure or mitigation of disease or injuries in humans by the administration of autologous, allogeneic or xenogeneic cells that have been manipulated or altered ex vivo."[30] The goal of cell therapy, overlapping with that of regenerative medicine, is to repair, replace or restore damaged tissues or organs. One way to do this is for the cells to integrate into the site of injury, replacing damaged tissue, and thus facilitate improved function of the organ or tissue. Another way for cells to be used is they can deliver treatment like a drug to a specific area. For our purposes what is of interest is the use of embryonic stem cells as treatment or therapy and the genetic modification of cells as a way to treat disease. According to Dr. West, here is how this might work.

One concrete example would be for the treatment of HIV infection. There are some rare individuals that have genetic resistance. We know the precise change they have in their DNA. We could then change the DNA in the embryonic stem cells that way and make the infected person a new immune system that could resist the virus. There is good evidence this could cure the disease.

Gene Therapy - Genes contain your DNA — the code that controls much of your body's form and function, from making you grow taller to regulating your body systems. Genes that don't work properly can cause disease. Gene therapy is the use of DNA as a drug to treat disease by delivering therapeutic DNA into a patient's cells. It can be done by:

---

[30] Application of Current Statutory Authorities to Human Somatic Cell Therapy Products and Gene Therapy Products, Notice, Oct 14, 1993 (Federal Register). (Notice from the Center for Biologics Evaluation and Research, Center for Drug Evaluation and Research, Center for Devices and Radiological Health, U.S. Food and Drug Administration.).

Replacing a mutated gene that causes disease with a healthy copy of the gene.

Inactivating, or "knocking out," a mutated gene that is functioning improperly.

Introducing a new gene into the body to help fight a disease.

There are two major types of gene therapy. The first is somatic gene therapy where any modifications and effects will be restricted to the individual patient only, and will not be inherited by the patient's offspring or later generations. The second type is germ line gene therapy. In germ line gene therapy, germ cells (sperm or eggs) are modified by the introduction of functional genes, which are integrated into their genomes, and in theory, are highly effective in counteracting genetic disorders and hereditary diseases. Some jurisdictions, including Australia, Canada, Germany, Israel, Switzerland, and the Netherlands[27] prohibit this for application in human beings, at least for the present, for technical and ethical reasons, including insufficient knowledge about possible risks to future generations[27] and higher risk than somatic gene therapy (e.g. using non-integrative vectors).[28] The USA has no federal legislation specifically addressing human germ line or somatic genetic modification (beyond the FDA testing regulations for therapies in general).[27][29][30][31]

From Wikipedia on Gene therapy

Potential ethical issues:

Gene Doping

There is a risk that athletes might abuse gene therapy technologies to improve their athletic performance.[87] This idea is known as gene doping and is as yet not known to be in use but a number of gene therapies have potential applications to athletic enhancement.

From Wikipedia on Gene Therapy

Human Genetic Engineering

It has been speculated that genetic engineering could be used to change physical appearance, metabolism, and even improve physical capabilities and mental faculties like memory and intelligence. These speculations have in turn led to ethical concerns and claims, including the belief that every fetus has an inherent right to remain genetically unmodified, the belief that parents hold the rights to modify their unborn offspring, and the belief that every child has the right to be born free from preventable diseases.[88][89][90] On the other hand, others have made claims that many people try to improve themselves already through diet, exercise, education, cosmetics, and plastic surgery and that accomplishing these goals through genetics could be more efficient and worthwhile.[91][92] This view sees the prevention of genetic diseases as a duty to humankind in preventing harm to future generations.[31]

From Wikipedia on Gene Therapy

---

[31] From Wikipedia on Gene Therapy.

The result of this technology is we can modify food, i.e. Genetically Modified Food (GMO), and are able to genetically modify insects, plants, animals and people. A biotech example is they took the gene that produces human insulin and they spliced it into bacteria, and suddenly these bacteria are a protein- or insulin-producing factory. That insulin is 100% human insulin, whereas the older insulin extracted from cows carried the danger of infectious viruses.

Animal models with human diseases can also be created, where the disease is genetic in nature. In other words, you can genetically modify the DNA of the animal so it has a human disease. This is much better for testing new drugs.

Another example, which is more "trivial" in nature is making transgenic art. In this example they took the gene that gives jellyfish their green glow and inserted it into the DMA of a rabbit. The rabbit as a result glows green under fluorescent light.

GFP Bunny -- With GFP Bunny Kac welcomes Alba, the green fluorescent rabbit, and explains that transgenic art must be created "with great care, with acknowledgment of the complex issues at the core of the work and, above all, with a commitment to respect, nurture, and love the life thus created." The first phase of the GFP Bunny project was completed in February 2000 with the birth of "Alba" in Jouy-en-Josas, France. The second phase is the ongoing debate, which started with the first public announcement of Alba's birth, made by Kac in the context of the Planet Work conference, in San Francisco, on May 14, 2000. The third phase will take place when the bunny comes home to Chicago, becoming part of Kac's family and living with him.

What is the significance of our being able to program genetic code? Let's say you have the letters C A T in a sentence. Well each letter means something in a word and each word means something in the sentence. Now, it is nothing for us to use a program such as Microsoft Word and change the "C" to a "B" and as a result change the word from CAT to BAT, which of course means something different. So when we read the sentence, "The CAT leaped off of the roof, it means something different than the bat leaped off of the roof. To continue the analogy, we can delete the word entirely, or insert a new world such as BLACK prior to the word CAT. We can change the word completely to panther. When we read the panther leaped off of the roof, still yet another meaning. We have the ability to change every word in the sentence. Just like in software, we can take a section of code that provides a functionality we desire and insert it in our code.

Now, if a scientist sees a sequence of code, say A T C C G G A, the scientist can change one letter in the sequence, delete the word, insert a new word, etc.

What are the implications of this? It now costs approximately $1,000 to have your genome mapped. The human genome is like the blueprint used for the design of your body. It is also like software code. Let's say you want children, you pay to have the genome of your child mapped at the pre implantation embryo stage. At that time you discover there is a sequence of code in your daughter's genome that has the propensity for breast cancer. A scientist could go in and cut out or delete that part of the code and potentially eliminate that problem. Or the scientist could fix the "bug" by inserting or changing some code.

Using this example, say once the genome in the pre implantation embryo is mapped and you get a report of all genetic diseases for which your daughter has over a 60% chance of getting in her lifetime. Perhaps the results report there are five such diseases and you can pay to have all five removed or just the ones with the highest priority.

One could imagine a scenario where life or medical insurance is less expensive for those who have this procedure done. All the person has to do is bring in the report of the genomic map study, list of which diseases were eliminated by genetic deletion or replacement and as a result, their life and medical insurance is less expensive. By removing the diseases the person has the highest propensity for getting, you have lowered the risk of them getting sick or dying. As a result, they are no longer the average person. Perhaps it would become the standard of care as all good parents would be seen as doing whatever they could to ensure their children lived a healthy life. In the same way we make the distinction between smokers and non-smokers, perhaps we raise the insurance on those who don't have the test done because they are at a much higher risk and thus most expensive to insure.

Some see this as acceptable ethical practice as you are still helping cure or treat disease.

However, what if you want to do more than to eliminate disease? What if you want to "enhance" what was there? In the above example, it is like moving from being a cat to being a cougar. This is a type of self-directed evolution. Some would argue that this is far more efficient than the evolutionary process, which by definition is hit or miss. With a large sample of genomes or code, we could select best in class for various functions.

In other words, the type of human modifications we want would be more like purchasing a car. What color do we want. . . select, cloth seats or the leather. . . select, HD stereo or regular, etc. So we select the enhanced hearing or eyesight, or intelligence, etc. Again, some might be comfortable with this level of genetic manipulation. What's wrong with bettering future generations?

The enhancements could be cosmetic. Perhaps we wanted to make sure our children had a nice brown skin tone, so they looked like they always had a tan and didn't have to worry about getting sun-burned or skin cancer. So we adjust the melanin to give them a certain type of skin tone. Say a father always wanted a son, which we could select, and wanted him to play in the NBA. We could tweak his genetic code to ensure he was at least seven feet tall. (What if the son has no interest in playing basketball? Being good in basketball requires more than height) Perhaps we want our child to have the best probability for entering college, so we replace his or her natural genetic code with code that has been highly correlated with intelligence. After all, if we are going to spend tens of thousands of dollars on his or her private school education, why not spend a few thousand to get this procedure done?

When we hear of these things, we often think of "designer babies." But what if it were put to us this way: What compassionate parent would knowingly let their child be born with genetic defects, when they could have prevented it? The problem in this case is this process requires IVF, which is expensive and not covered by most insurance plans. So there might quickly become two segments of people, separated by social class. Those that can afford this type of medical procedure and those who can't.

Genetic programming/genetic modification can go ever farther. By using "transgenetic" techniques, we can splice in codes from other species such as insects or animals.

What if we wanted our son to be a star athlete or have an edge versus the other kids? Perhaps we think it is better to pay for this work now in the hopes he gets a scholarship to college later. So we find a characteristic or the genetic code responsible for lean body mass in a certain type of animal and we replace the code used in our son's code with it. In other words it is said that ants can lift several times their body weight. Perhaps there is an animal that has a similar characteristic. So we insert that code into our son's genetic code so he will have exceptional strength relative to his body mass.

What if we could create the code for programs that don't exist anywhere in nature? By knowing the API for smartphones, billons of programs have been created and something similar could be done for the human API. This would enable us to do amazing things that probably would never have occurred through the process of natural selection. Earlier in the discussion we talked about mixing and matching genes from different species (plants, insects, animals and people). Here we are talking about creating entirely new code to provide us capabilities that don't exist in nature. With the computing power we have today, we could try millions upon millions of genetic sequences in an effort to develop new code, programs (characteristics) that could provide us the ability to do things that have never been done in nature by any insect, plant, animal or person.

Imagine creating species where you can program precisely, and mix and match characteristics from among plants, insects, animals and people. We can now treat genes like a large set of Lego blocks and put them together anyway we want. *We have the ability to "program" the DNA to create new combinations of life forms that have never existed before.* We have already seen the potential danger of introducing non-native species into an environment, killer bees, ants and fish are all familiar examples. What could the ecological impact be with transgenetic animals. . . or people?

From an evolutionary perspective, species change over time. With the development of breeding technology, we were able to breed for the characteristics we desired. The new technology which would let us perform "Genetic Programing" or Genetic Modification is a quantum leap in breeding.

- Breeding is still somewhat random.
- Breeding is limited to within a species
- Breeding is very imprecise in some instances

We now have the ability to do with humans what we could not have dreamed of with animals and we can do it faster and more accurately.

Nuclear Transfer Technology

Somatic Cellular Nuclear Technology (SCNT) – more commonly referred to as cloning – makes it possible to apply this technology to humans. Given we now have the human genome and various gene targeting techniques as described, scientists have the ability to modify human DNA and then to use nuclear transfer technology not to create a clone but to make it possible for this modified cell to start diving and become an embryo which can be implanted into a womb.

Rather than creating an exact replica, you are creating a genetically modified human.

What are the potential applications of genetic programing/genetic modification in humans, also referred to as germ line modification or generic modification?

Physical Characteristics- e.g. height, weight, size, texture, gender, etc.

Personality – (as much as it is determined by genetics) attitude, emotions, IQ.

Enhancements:
- Eye color versus vision improvement. Beyond splicing in the genes of someone that has better vision, trying to transfer the gene that provides a hawk its keen eye sight.
- Hearing, adding sonar and radar characteristics
- Muscle Mass – Knock-out GDF-8 which controls muscle mass in mice and cows? Imagine a zealous parent who wanted his son to be a football player.

Disease Resistance
- *Viral* - Improve the resistance to certain types of viruses, mumps, measles, AIDS
- *Bacterial* – improve the resistance to certain types of bacterial diseases

Genetic Defects
- Defects due to bad genetics. There are 3,000 known gene specific diseases. These could all be eliminated. Why install bad or defective code in your child?

The question is no longer what can we do, but what should we do?

When it comes to genetics, we often turn to eugenics and the experience in Nazi Germany. However, that fear is somewhat misguided. Most likely, someone's DNA will be "hacked." As the procedure becomes easier to perform, more people have the capability to genetically modify human DNA. Imagine if James Eagan Holmes, the Ph.D. candidate in neuroscience, also known as the Aurora Colorado shooter, worked in an IVF clinic and wanted to "hack" the DNA of a donor egg or the DNA of an embryo itself. No one would know until it was too late. Imagine if hacking was as prevalent in biology as it is in computer science.

Josef Mengele a Nazi war doctor, had doctorates in anthropology and medicine and was a practicing physician. Look at what he did in the name of science. Dr. Brigitte Boisselier of Clonaid made it known she wanted to clone a human being, even though the vast majority of the population is against such a thing. It is not the crazy people whom we need to fear; instead, we ought to fear those who have a skewed sense of what is ethical.

Another scenario might be a country such as China that has the resources and interest in achieving national pride does something that makes the world take notice. Given how prevalent doping scandals are in sports, imagine if a country wanted to produce a winning Olympic team and created 100 seven-foot tall men with the hope they could produce the world's best basketball or volleyball team, or something similar, in an effort to win a gold medal and bring pride to the country. It is easy enough to get surrogates in the US; imagine the motivation to be a surrogate in another country when the government is providing the incentive. Or consider very affluent individuals who want to have a child with certain characteristics and are willing to pay whatever it costs to get them. Typically, rules are not universal, so even if it were illegal in the U.S. to perform a particular procedure, it would probably be legal somewhere in the world. That's why people go to other countries to get access to drugs not approved in the U.S. If the "experiment" was successful, it would put more pressure on other countries to follow.

As a result, we will more likely than not incrementally move in this direction of genetic modification unless we make a decision not to.

Developing the Ethical Framework

Some could argue a position for no genetic modification of any kind. You could argue you can only use genetic modification to remove or fight a known disease. Another option is to limit genetic modification to selecting or inserting best in class features. There are numerous possibilities, which is why I believe we need a national discussion on this topic. As a starting point we can look at five options.

Our first option is to use this technology in the diagnosis and treatment of disease as long as it does not result in a change in the genetic structure of germ line.

The second option is called passive genetic modification. This type of modification involves looking at the embryos and only selecting the ones that do not have a (particular) genetic defect. If there is a strong belief a couple has a high probability of having a child with a genetic problem, that gene could be screened prior to implantation. This is Pre-implantation Genetic Analysis or Diagnosis.[32]

---

[32] The Ethics of Inheritable Genetic Modification: A Dividing Line? Edited by John Rasko, Gabrielle O'Sullivan, Rachel Ankey.

The third option is to attempt to eliminate "defects" and disease. If an embryo has a genetic code for a particular type of disease, it would be ok to eliminate that defect.

The fourth option would be to allow enhancements. It would be ok to enhance humans with specific improvements that existed in other humans.

The fifth option would be to allow enhancements from other species and permit the development of transgenetic humans.

There are other options. The point is we need to have the discussion because we need rules and regulations. Not every scientist has the same sense of right and wrong, and, with so much money at stake, one's values can get skewed. Now is the time to take a leadership position in this area and provide the world's scientific community recommendations or guidelines. That will allow us to use these new technologies to harness the promise and eliminate the peril.

I believe we should have more input than this notice on the FDA web site.

February 25-26, 2014: Cellular, Tissue, and Gene Therapies Advisory Committee Meeting: Announcement[33]

Center	Date	Time	Location
CBER	February 25, 2014	8 a.m. - 5:30 p.m. 8 a.m. - 5 p.m.	Hilton Washington, D.C.

---

[33] http://www.fda.gov/advisorycommittees/calendar/ucm380042.htm.

February 26, 2014		North/Gaithersburg, 620 Perry Pkwy., Grand Ballroom, Gaithersburg, MD 20877 (301-977-8900)

Agenda

On February 25, 2014, from 8 a.m. to 5:30 p.m. and on February 26, 2014, from 8 a.m. to approximately 11:15 a.m., the committee will discuss oocyte modification in assisted reproduction for the prevention of transmission of mitochondrial disease or treatment of infertility. On February 26, 2014, from approximately 11:15 a.m. to 11:30 a.m., the committee will hear updates on guidance documents issued from the Office of Cellular, Tissue, and Gene Therapies, Center for Biologics Evaluation and Research (CBER), FDA. On February 26, 2014, from 1 p.m. to approximately 5 p.m., the committee will discuss considerations for the design of early-phase clinical trials of cellular and gene therapy products. CBER published guidance on this topic in July 2013.

Meeting Materials

Materials for this meeting will be available at the Cellular, Tissue, and Gene Therapies Advisory Committee Meeting main page.

Public Participation Information

Interested persons may present data, information, or views, orally or in writing, on issues pending before the committee.

- Written submissions may be made to the contact person on or before February 18, 2014.
- Oral presentations from the public will be scheduled between approximately 2:15 p.m. and 3:15 p.m. on February 25, 2014 and between approximately 1:45 p.m. and 2:15 p.m. on February 26, 2014. Those individuals interested in making formal oral presentations should

notify the contact person and submit a brief statement of the general nature of the evidence or arguments they wish to present, the names and addresses of proposed participants, and an indication of the approximate time requested to make their presentation on or before February 10, 2014. Time allotted for each presentation may be limited. If the number of registrants requesting to speak is greater than can be reasonably accommodated during the scheduled open public hearing session, FDA may conduct a lottery to determine the speakers for the scheduled open public hearing session. The contact person will notify interested persons regarding their request to speak by February 11, 2014.

- For those unable to attend in person, the meeting will also be Webcast. The link for the Webcast is available at:

  - DA intends to make background material available to the public no later than 2 business days before the meeting. If FDA is unable to post the background material on its Web site prior to the meeting, the background material will be made publicly available at the location of the advisory committee meeting, and the background material will be posted on FDA's Web site after the meeting.

  - A notice in the Federal Register about last minute modifications that impact a previously announced advisory committee meeting cannot always be published quickly enough to provide timely notice. Therefore, you should always check the agency's Web site and call the appropriate advisory committee hot line/phone line to learn about possible modifications before coming to the meeting.

- Persons attending FDA's advisory committee meetings are advised that the agency is not responsible for providing access to electrical outlets. FDA welcomes the attendance of the public at its advisory committee meetings and will make every effort to accommodate persons with physical disabilities or special needs. If you require special accommodations due to a disability, please contact Gail Dapolito at least 7 days in advance of the meeting. FDA is committed to the orderly conduct of its advisory committee meetings. Please visit our Web site for <u>procedures on public conduct during advisory committee meetings</u>.
- Notice of this meeting is given under the Federal Advisory Committee Act (5 U.S.C. app. 2). <u>Official FR Notice</u>

The F.D.A. advisory panel says its meeting only considered scientific aspects of mitochondrial manipulation[34] and any "ethical and social policy issues" are outside its scope. But those are precisely the issues that we must address.[35]

Just as we don't know the side effects of a drug, which is why it goes through clinical trials, turning genes "on and off" and splicing genes in and out could have side effects

---

[34] "The procedures involve removing the nuclear material either from the egg or embryo of a woman with inheritable mitochondrial disease and inserting it into a healthy egg or embryo of a donor whose own nuclear material has been discarded," writes Marcy Darnovsky in a recent NYT editorial. "Any offspring would carry genetic material from three people -- the nuclear DNA of the mother and father, and the mitochondrial DNA of the donor."

[35] Marcy Darnovsky, Genetically Modified Babies, New York Times February 23, 2014.

we would only know about after the fact. These side effects have occurred in mice, where making an alteration to a specific gene had unintended consequences. Should these experiments be allowed, even if the patients are willing? Who speaks for the embryo, not in terms of life or death, but what type of modification experiment is permissible because any defect may only be determined long after we have a human living life on the planet.

I wrote to Senators Elizabeth Warren and Ed Markey. This is what my letters said in part:

> I am asking you to ask President Obama to have his Bioethics Commission, or a new one to be named, to recommend ethical guidelines for genetic modification[36] as it relates to humans. Earlier this year you may have read headlines about the first genetically modified babies born.[37] Scientists may tell you what was done was not technically genetic modification, even though the babies that resulted had two sets of female genes and one set of male genes. What they did do I believe would be alarming to most Americans. I believe we should have some national discussion on this topic before a scientist somewhere does something we believe is unacceptable as a nation. It is especially important as several institutions involved in researching these topics are in Massachusetts - Whitehead Institute for Biomedical Research and Harvard University

---

[36] Also referred to as genetic engineering, germ line modification and genetic programming. It is not clear we have dealt with the ethics of gene therapy and personalized medicine based on an individual's genome and this goes beyond that by an order on magnitude.
[37] Genetically Modified Babies, Marcy, Darnovsky New York Time February 23, 2014, Report First Genetically modified babies, ABC News May 4, 2014.

Department of Stem Cell and Regenerative Biology. Advanced Cell Technology is one of the leading companies in this field and it is a Massachusetts company. It is crucial that thoughtful policies are developed to meet the needs of medical science and our shared ethical values.

There are four major technologies, genomics, genetic engineering, embryonic stem cells, and nuclear transfer (cloning) that independently have great promise and peril for us. When combined they are powerful in terms of regenerative medicine and the genetic modification of humans. So we need limits. The four technologies mentioned provide us the tools to turn insect, plant, animal and human genes into "Tinker Toys" that can be arranged in anyway desired, allowed or "hacked" in the same way we hack toys or computer code.

Genetic modification could be used to treat diseases, enhance human characteristics, or provide humans characteristics that have never existed before. While the creation of "Green Rabbits" is considered art, the same results could be achieved with humans. I wrote a small book on this topic called "Regenerative Medicine and Human Genetic Modification", http://amzn.to/1mnPCy0 and in it is a chapter called "No Genetic Modification of Food or Humans?" I argue we should have at least as much discussion on the modification of humans as we do on food e.g. the labeling of genetically modified food. If you go to Amazon.com and search on "regenerative medicine" or "genetic modification", you will see few books there on the topic, how expensive they are and how dated most of them are. Something of this magnitude should have a much broader discussion.

Because these new developments (regenerative medicine and human genetic modification) not only advance medical science, but are patentable, potentially billions of dollars will be made with new discoveries that not only treat disease but change human characteristics. Each new genetic code to enhance life would be worth millions and foster a gold rush as people try to identify new ways to enhance humans, similar to developing apps for cells phones found at the iTunes store. Given the amount of potential money to be made, industry is likely to have some influence on our ethics as we try to determine what is right and wrong. Therefore, there is some need to get ahead of this issue. We should not trust industry to do the right things when millions of dollars are on the table.

Perhaps this sounds more like science fiction than science, but consider that the study of life science is increasingly about genetics, and genetics is becoming a question of computing power, machine learning and "big data." That plays into Google's strength. "Digital insights are becoming increasingly important in the life sciences, because genetics is a data-driven practice, giving us a real place to add value," says Dr. Krishna Yeshwant, a general partner with Google Ventures.[38] Google has been increasing its investments in the life sciences. Google has a life science startup called Calico that will pursue solutions for aging and its associated diseases. Calico, with its focus on anti-aging or "radical life extension," will intersect with the work done by those in the regenerative medicine community. This Google division is making other investments in the life sciences and has a team of between 70 – 100 full-time scientists.

In addition to investments, Google was able to get Art Levinson, the former CEO and current chairman of Genentech involved. Genentech was one of the pioneers in genetic engineering. Google recently launched a project it called "Baseline Study"[39]. The project will collect genetic and molecular information to create the fullest picture of what a healthy human being should look like. The idea is, in order to fix something, you need to know what "well" or "healthy" looks like as a baseline. In terms of diagnostics, biomarkers could be developed that would help detect the

---

[38] Understanding Calico: Larry Page, Google Ventures, and the quest for immortality, The Verge, Ben Popper, September 19, 2013 01:10 pm.
[39] Google's New Moonshot Project: the Human Body, Baseline Study to Try to Create Picture From the Project's Findings, Alistar Barr, Wall Street Journal July 27th.

onset of chronic diseases much sooner. The hope is this will help researchers detect killers such as heart disease and cancer far earlier, pushing medicine more toward prevention rather than the treatment of illness.

To summarize, think of it this way, we can map everyone's human genome for approximately $1,000. That enables us to develop personalized medicine designed specifically for that person's genome. As a result, personalized therapies will work better than drugs or therapies that are designed for the general population. (That's why some drugs work better than others when trying to treat a patient.) While we have the genetic map, and are able to spot "bugs" or defects in the genetic code, we don't know what is "ideal." Google hopes the work it is doing will tell us. If we know what the idea is, we have the ability to modify it using the tools of genetic engineering, embryonic stem cells, therapeutic cloning, and genetic engineering, the tools of genetic modification.

What we have argued in this book is knowing what "healthy" looked like, Google could provide the necessary insight to change the genetic code while it is still in the pre embryonic stage and help prevent or reduce over 3,000 genetic diseases. By knowing what "healthy" looks like, they would also be in a position to identify outliers. Genetic code that was superior to the average would be identified and its information made available, perhaps for a cost for those who wanted to enhance the genetic structure of their child. Through computer modeling, they could develop genetic code they predict will have a benefit and perhaps make that available for sale as well. What if Google looked at the genetic code of the 1,000 smartest people in the world, discovered patterns they believed were the drivers of intelligence and made those genetic maps available for a price. How much would people be willing to pay to give their son or daughter the opportunity to be one of the smartest people in the world?

Google sees as part of its mission organizing the world's information and genetic data and other health science data is part of that mission. Google is not the only such company that can play in this area and more will enter as this is estimated to be a multi-trillion dollar market.

Perhaps you have similar feelings and desire to take action. At least you are now aware of the problem. Where we go from here depends upon what individuals such as you decide to do. The next step is up to you. As you can see from the bibliography below, now is the time to act. Write your Congressperson and ask for a hearing, start a petition on change.org. With regenerative medicine and human genetic modification, we have the ability to treat disease; chronic, genetic and infectious, enhance human characteristics or even create new human organisms. We have to make a collective choice about the future of the species.

Ananda, Rady. "Genetically Modified Babies." *Global Research*. N.p., 15 Oct. 2013. Web.

Arnold, Paul. "Pros and Cons of Genetic Engineering in Humans." *Bright Hub*. N.p., 28 Feb. 2012. Web. <http://www.brighthub.com/science/genetics/articles/22210.aspx>.

Darnovsky, Marcy. "Genetically Modified Babies." *The New York Times*. The New York Times, 23 Feb. 2014. Web.

DiSalvo, David. "The Era Of Genetically-Altered Humans Could Begin This Year." *Forbes* (2014): n. pag. Web.

Gropp, Michal, Pavel Itsykson, Orna Singer, Tamir Ben-Hur, Etti Reinhartz, Eithan Galun, and Benjamin E. Reubinoff. "Stable Genetic Modification of Human Embryonic Stem Cells by Lentiviral Vectors." *Molecular Therapy* 10.1016/S1525-0016(02)00047-3 (2003): 281-87. Print.

Hanlon, Michael. "World's First GM Babies Born." *Mail Online*. Associated Newspapers, n.d. Web.

Huff, Ethan A. "Genetically-modified Humans Are Already Walking among Us." *NaturalNews*. N.p., 15 May 2013. Web.
<http://www.naturalnews.com/040348_gm_babies_fda_human_genome.html>.

"Human Genetic Engineering, Current Science and Ethical Implications Fact Sheet." Council for Responsible Genetics, n.d. Web.
<www.councilforresponsiblegenetics.org>.

Mercola, Joseph. "Will These Genetically Modified Babies Alter Human Species?" *Mercola.com*. N.p., 17 July 2012. Web.
<http://articles.mercola.com/sites/articles/archive/2012/07/17/first-genetically-modified-babies-born.aspx>.

Simmons, Danielle, Ph.D. "Genetic Inequality: Human Genetic Engineering." *Nature Education* 1.1 (2008): 173. Print.

Author's Note

When I was telling my friends about this book, people heard the title and assumed that the book was over their heads, it wasn't. But when I told people about my fiction based on these concepts, people loved them. Here are a couple examples.

Night Vision – A woman takes her child to get an eye exam. The bulb goes out in the machine but the child can still perfectly read the eye chart in the dark. Why, how? Unbeknownst to the mother, the doctor at the IVF clinic she used had inserted the genetic code that gives nocturnal animals eyesight, into the DNA of her embryo, which is why the child had "Night Vision."

UnDesigned – the world is divided into two types of people, those who are designed, as in designer babies and those who are not. Because genetic modification is not covered by insurance, those who are designed are thought to be the elite, the upper class. The series revolves around an attorney, Jenn who works for a high powered law firm who has clients who are in the genetic modification industry and as such are always being sued.

I have written a number of genetics based science fiction. If these are of interest to you, send me an email at ewgaskin@gmail.com. I may publish them as a collection under the "Night Vision" title.